# Creating Documents

# with

# Scientific WorkPlace™

# and

# Scientific Word®

# Creating Documents

## with

## Scientific WorkPlace™

## and

## Scientific Word®

Roger Hunter
*TCI Software Research*

Susan Bagby
*TCI Software Research*

SOFTWARE RESEARCH          TOOLS FOR SCIENTIFIC CREATIVITY

TCI Software Research
Las Cruces, New Mexico

Brooks/Cole Publishing Company
Pacific Grove, California

**TCI Software Research**

A Division of Brooks/Cole Publishing Company

Printed in the United States of America

10  9  8  7  6  5  4

**Trademarks**

*Scientific Word* is a registered trademark of TCI Software Research. *Scientific WorkPlace* is a trademark of TCI Software Research. EasyMath is a registered trademark of TCI Software Research. EasyMath is the sophisticated parsing and translating system included in Scientific WorkPlace that allows the user to work in standard mathematical notation, request computations from the underlying computational system (Maple in this version) based on the implied commands embedded in the mathematical syntax or via menu, and receive the response in typeset standard notation or graphic form in the current document. Maple is a registered trademark of Waterloo Maple Inc. TeX is a trademark of the American Mathematical Society. TrueTeX is a registered trademark of Richard J. Kinch. Windows is a registered trademark of Microsoft Corporation. All other brand and product names are trademarks of their respective companies. The spelling portion of this product is based on Proximity Linguistic Technology. Words are checked against one or more of the following Proximity Linguibase® products:

| Linguibase Name | Publisher | Number of Words | Proximity Copyright |
|---|---|---|---|
| American English | Merriam-Webster, Inc. | 144,000 | 1990 |
| British English | William Collins Sons & Co. Ltd. | 80,000 | 1990 |
| Catalan | Lluis de Yzaguirre i Maura | 484,000 | 1992 |
| Danish | Munksgaard Int. Publishers Ltd. | 169,000 | 1990 |
| Dutch | Van Dale Lexicografie bv | 223,000 | 1990 |
| Finnish | IDE a.s | 191,000 | 1991 |
| French | Hatchette | 288,909 | 1992 |
| French Canadian | Hatchette | 288,909 | 1992 |
| German | Text & Satz Datentechnik | 500,000 | 1991 |
| German (Swiss) | Text & Satz | 500,000 | 1992 |
| Italian | William Collins Sons & Co. Ltd. | 185,000 | 1990 |
| Norwegian (Bokmal) | IDE a.s | 150,000 | 1990 |
| Norwegian (Nynorsk) | IDE a.s | 145,000 | 1990 |
| Portuguese (Brazilian) | William Collins Sons & Co. Ltd. | 210,000 | 1990 |
| Portuguese (Continental) | William Collins Sons & Co. Ltd. | 218,000 | 1990 |
| Spanish | William Collins Sons & Co. Ltd. | 215,000 | 1990 |
| Swedish | IDE a.s | 900,000 | 1990 |

This document was produced with *Scientific WorkPlace*.

Sponsoring Editor: *Robert Evans*
Editorial Assistant: *Tami McBroom*
Production Coordinator: *Marlene Thom*
Manuscript Editor: *Harriet Serenkin*
Cover Design: *Robert Western*
Compositor: *TCI Software Research*
Printing and Binding: *Malloy Lithographing, Inc.*

# Contents

## 11   Structuring Documents     159

# What's New in the Workplace

Welcome to the Workplace: The place where you create complex technical documents with ease. The place where you use natural notation, not special codes and equation editors, to create and display mathematics and where you can perform mathematical computations right in your document. The place where you produce beautiful, typeset-quality documents. With *Scientific WorkPlace* and *Scientific Word*, you can create mathematics and text on the same screen and can produce typeset-quality formulas, equations, and documents. And with *Scientific WorkPlace*, you can perform a wide range of mathematical computations using a special version of Maple, the symbolic computation system, or a link to your Mathematica software. Both *Scientific WorkPlace* and *Scientific Word* incorporate an easy-to-learn, easy-to-use scientific word processor.

This version of *Scientific WorkPlace* and *Scientific Word*, which we refer to together as *SW*, makes creating technical documents even easier. With 32-bit processing, multiple windows and views, an improved interface, a customizable screen, and user-definable defaults, you can create a workplace that's tailored to the way you work. Hypertext links, external program calls, verbatim input, and increased support for non-English languages greatly expand the possibilities for the content of your documents. With the completely redesigned Style Editor you can create printed documents that meet your formatting requirements. And getting help when you need it has never been easier.

These new features add to those that characterized earlier versions of the software. They enhance your ability to create and edit your content, perform symbolic computations, and produce beautiful, typeset-quality documents.

## Basic Features

If you used an earlier version of *SW*, you'll recognize these important features in version 2.5.

### The SW Approach

The heart of the *SW* approach to word processing is the separation of content and appearance. The content of your work results from the creative process of forming ideas and putting them into words. The appearance of your work results from typesetting, the mechanical process of displaying the document on the printed page in the most readable format.

The *SW* approach is known as *logical design*. It separates the creative process of writing from the mechanical process of typesetting. Logical design frees you to focus on creating the content instead of the format, increases your productivity, and leads to a more consistent and higher-quality document appearance because the format is applied automatically. Separating the processes of creating and formatting a document combines the best of the online and print worlds. You create a good document; *SW* creates a beautiful one.

## Rich Interface

- **Apply what you already know** about using Windows applications directly to the *SW* environment. First, you'll find the available commands in menus, bars, buttons, and popup lists on the screen. Many commands have associated dialog boxes, as indicated by the ellipses that follow the menu commands and buttons. Second, you can enter commands from the keyboard or with the mouse using standard keyboard and mouse techniques. You can speed your work with drag-and-drop editing and speed scrolling. Third, you'll recognize many standard Windows keyboard conventions, such as using CTRL+X to delete and SHIFT+ARROW KEYS to extend a selection. For a review of these conventions, see your Windows documentation.

- **Use a comprehensive set of keyboard shortcuts and key prefixes**—special key combinations for performing basic operations and entering symbols, characters, and the most common mathematical objects. The keyboard shortcuts and key prefixes are often faster to use than the mouse.

- **Work with pleasing on-screen mathematics and italics** created with TrueType and Type 1 outline fonts. *SW* uses Windows 3.1 (or higher) TrueType fonts. Some of the required fonts are supplied with Windows, and some are supplied with *SW*. You can also use Adobe Type Manager and Adobe Type 1 fonts.

## Natural Entry of Text and Mathematics

- **Enter text and mathematics** in the same paragraph. The *SW* screen default is to show text in black and mathematics in red so you can easily distinguish them. The Math/Text button on the Math toolbar indicates whether the insertion point is in text or in mathematics:

   The insertion point is in text.

   The insertion point is in mathematics.

  Everything you enter is assumed to be text unless you specify otherwise. You can toggle between text and mathematics easily with the mouse or the keyboard.

- **Use templates for entering mathematical objects** such as fractions, radicals, enclosures, and matrices by inserting and filling the template for the object you want. For example, if you insert a fraction, *SW* places this template into your document:

  and places the insertion point in the small input box that appears in the numerator so you can complete the fraction. You do not need templates to enter individual mathematical symbols and characters. You enter them directly into your expression using the menus, buttons, or key prefixes.

- **Define automatic substitution sequences** to speed the entry of the mathematical expressions you use most often. When the insertion point is in mathematics and you type the substitution sequence, *SW* replaces the sequence with the corresponding expression. You can turn automatic substitution off or on at any time.

- **Use the full set of American Mathematical Society symbols.** Incorporate in your documents all the symbols in the AMS fonts. *SW* supports the full set of AMS fonts, both on the screen and in print.

- **Create multiline displays quickly** and add numbers selectively to each line of the display for easy cross-referencing throughout your document. You can override the automatically generated numbers with your own labels.

- **Place text and mathematics in tables** with the easy-to-use table editor.

## Mathematical Computations Inside Your Document

Perform mathematical computations directly from the main window of *Scientific Work-Place*:

- **Execute implied operations.** Apply Evaluate to display the result of operations implicit in the document, from basic arithmetic operations to computations in calculus and linear algebra.

- **Apply mathematical operations.** The Maple menu in *Scientific WorkPlace* provides many operations you can apply to solve mathematical problems. Solve systems of linear equations or differential equations, fit curves to data, and experiment with graphs.

- **Obtain mathematical results separately or in place.** Select a piece of mathematics and apply an operation to obtain the result directly in your document. Press the CTRL key during the operation to replace the original expression with the result of the computation.

- **Create graphics.** Create two- and three-dimensional plots with Plot 2D and Plot 3D Rectangular, or use the many other options provided on the Plot 2D and Plot 3D submenus. Use convenient plot tools to zoom and pan two-dimensional plots and rotate three-dimensional plots to obtain optimal views. Use drag-and-drop mouse techniques to add new functions to a graph quickly and easily.

- **Create and operate on matrices.** Create matrices with Insert Matrix or Fill Matrix. Carry out implied operations simply by applying Evaluate to a sum, product, or power. Apply the numerous mathematical operations displayed on the Matrices submenu.

- **Operate with the names of functions**, after defining them by expressions. Define generic functions and use them to illustrate rules for differentiation or use the names of generic functions in defining new functions.

- **Make use of user-defined Maple functions.** Use the Define Maple Name dialog box to access and take advantage of functions that have been programmed in full Maple.

## Easy Creation of Complex Documents

- **Use tags to add structural elements** such as lists, sections, and theorems to your document and to emphasize words and phrases in the text. In *SW*, each paragraph has an associated tag that determines whether it is a section heading or part of the body of your document.

- **Manage large documents with ease** by creating a master document that incorporates several smaller, more manageable subdocuments. When you print the master document, *SW* creates all surrounding front and back matter specified by the typesetting style and resolves all cross-references internal to all subdocuments.

- **Import files directly into an active document** to speed document creation. Use the Import Contents feature to copy the contents of other files into your active document. *SW* imports the body of the document but not the style or back matter.

- **Illustrate your ideas with graphics** created in popular software applications and imported to *SW*. Enhance your documents with graphics generated in TIFF, PICT, WMF, BMP, DIB, CGM, EPS, and many other formats.

- **Create indexes and bibliographies easily.** Simplify the generation of back matter for your document with a streamlined procedure for creating indexes and the ability to create either manual or BIBTEX bibliographies.

- **Resolve all document cross-references automatically.** Create cross-references to sections, pages, and other markers in your document, and then let *SW* and LATEX do the work of resolving the references when you compile your document.

## Beautiful Printed Documents

- **Produce typeset-quality printed documents automatically.** Your documents are typeset using LATEX, a set of macros designed by Leslie Lamport to enhance TEX with document-structuring features such as tables of contents and bibliographies. TEX is the extraordinary program and language for mathematics typesetting designed by Donald Knuth. The Windows implementation of TEX and LATEX supplied with *SW* is TrueTEX, a product of Kinch Computer Company. Because typesetting is automatic, you do not have to know TEX or LATEX to use *SW* effectively.

- **Determine the appearance of your document by choosing from a wide selection of predefined typesetting styles.** Select the typeset appearance of your document from more than 100 typesetting styles for producing books, articles, reports, and other types of documents. Because the styles contain complete document-formatting instructions, you can concentrate on writing your document instead of on formatting its appearance.

- **Define new typesetting styles with the Style Editor.** Use the Style Editor to create special typesetting styles that match your formatting requirements; then save those styles for future use. You don't have to know LATEX to create typesetting styles with the Style Editor (not available in the Student Edition).

- **Create AMS-LATEX documents.** *SW* now creates documents in $\mathcal{AMS}$-LATEX format automatically when you choose one of the $\mathcal{AMS}$-LATEX typesetting styles.

- **Produce documents using REVTEX,** a package of LATEX macros designed for preparing physics manuscripts. Version 2.5 includes several styles for producing documents in REVTEX formats.

- **Take advantage of the latest in TEXnology.** Use the latest versions of TEX and LATEX to create documents with a much wider variety of scalable fonts. *SW* includes support for the New Font Selection Scheme, LATEX $2_\varepsilon$, and a high-capacity TEX that processes your documents quickly. Your *SW* package also includes TrueTEX, featuring a previewer that uses TrueType and Type 1 fonts, supports a wide range of graphics formats, and enables you to print on any Windows-supported device including fax boards.

- **Preview the typeset appearance of your documents with more than one previewer.** Choose from all installed previewers when you're ready to see your document as it will appear in print. You can also use the Compile feature to create a typeset file without previewing or printing until you're ready to see the typeset document.

## New Features

The following features are new in version 2.5 of *SW*. More information on each feature is available in this user's guide, in online Help, and in the document readme.tex in the docs directory.

- **Take advantage of improved performance with 32-bit processing.** Create and produce documents faster with built-in 32-bit processing. Use *SW* in the 32-bit environments of Windows 3.1 with Win32s 1.3, Windows 95, and Windows NT to write, preview, and print your documents.

- **Use enhanced computational capabilities.** Perform mathematical computations even more intuitively with substitution in expressions, deferred evaluation, subscript variables, and other enhancements to the Maple interface. If you use Mathematica, *SW* now offers a connection to your Mathematica software.

- **Open several windows at once.** Open several *SW* documents and several views of the same document at the same time. The changes you make in one view of a document are recorded in all views of the same document.

- **Arrange the screen for your convenience.** Dock the toolbars where you want them on the top, bottom, or sides of the *SW* main window; hide those toolbars you don't need; and display the symbol panels while you work. *SW* remembers the toolbar and panel placement from session to session.

- **Speed your work with expanded toolbar operations.** Use the buttons on several new toolbars to speed up standard file and editing operations and basic mathematical calculations. The buttons mirror commands on the File, Edit, Insert, Go, and Maple menus. Tooltips identify the associated menu commands.

- **Develop your own typesetting styles with the new Style Editor.** Set the specifications for fonts, paragraphs, and other elements of style more intuitively and with greater flexibility, and preview your style from the completely redesigned Style Editor. (The Style Editor is not available in the Student Edition.) When all you need is a simple document, create letters, memos, and reports with the new Easy Styles provided with *SW*.

- **Create dynamic documents with hypertext links.** Use the links to jump to paragraphs, sections, or objects with an identifying key, then retrace your steps with the history feature. The ease of entering cross-references has been improved in *SW* version 2.5 with drop-down lists of defined markers.

- **Insert external program calls into your document.** Call demonstrations, interactive tutorials, and other programs from within your *SW* document.

- **Use SW with Western European and some Eastern European languages.** *SW* now provides fonts and hyphenation support for languages other than English. You can switch languages in the same document by using Babel, the multilingual LaTeX system, and you can enter your documents from non-U.S. keyboards.

- **Create and edit documents more easily with an improved interface.** Use both text and mathematics in the lead-in objects for lists and theorems. Enter spaces, rules, fills, and breaks directly from dialog boxes, and specify tags for standard LaTeX text sizes such as tiny, large, and huge.

- **Find more help online.** Use the expanded online *SW* Help feature to get help when you need it. The *SW* manuals are available through the Help menu, as is a series of tutorial exercises designed to help you learn to use the most important features of the software. The new Style Editor has its own fully illustrated and context-sensitive online Help feature.

- **Load documents quickly with the Quick Load feature.** Speed document loading by turning on the *SW* Quick Load feature. This new feature is especially useful for large documents used as interactive texts.

- **Set the defaults for your workplace.** Use the redesigned User Setup dialog box to set the defaults for measurement units, mouse button uses, start-up styles, placement and size of new graphics, automatic saving, and many other aspects of the *SW* environment.

- **Display program code in SW.** Use the new verbatim environment to include portions of programs in your document.

## About this Manual

This manual is your guide to working with *SW*. We've organized the manual by tasks, beginning with the most basic word-processing tasks—those things you need to do for nearly every document you write—and moving to more advanced tasks.

## Before You Begin

You must install Microsoft Windows 3.1 (and Win32s 1.3), Windows NT, or Windows 95 on your computer before you install version 2.5 of *SW*. Your Windows documentation contains information on installing Windows. The booklet *Getting Started with Scientific WorkPlace and Scientific Word* contains instructions for installing *SW* on an individual computer or a network.

## Conventions

Understanding the notation and the terms we use will help you understand the instructions in this manual.

### General Notation

- *Text like this* indicates important terms introduced in the text.
- **Text like this** indicates text you should type exactly as it is shown.
- ***Text like this*** is a placeholder for a file name or a directory name that you must supply.
- Text like this indicates the name of a menu, command, or tag.
- TEXT LIKE THIS indicates the name of a keyboard key.
- Text like this indicates the name of a file or directory.
- The notation *SW* means both *Scientific WorkPlace* and *Scientific Word*.
- The word *choose* means to designate a command for the program to carry out. The command may be listed on a menu or shown on a button in a dialog box. As with all Windows applications, you can choose a command with the mouse or with the keyboard. For example, the instruction "From the File menu, choose Open" means you should first choose the File menu and then from that menu, choose the Open command. The instruction "choose OK" means to click the OK button with the mouse or press TAB to move the attention to the OK button and then press the ENTER key on the keyboard.
- The word *select* means to highlight the part of the document that you want your next action to affect or to highlight a specific option in a dialog box or list.
- The word *check* means to turn on an option in a dialog box.

### Mouse Conventions

In this manual we give mouse instructions using standard Windows conventions.

- *Point* means to move the mouse pointer to a specific position.
- *Click* means to position the mouse pointer, then press and immediately release the left mouse button without moving the mouse.
- *Double-click* means to position the mouse pointer, then click the left mouse button twice in rapid succession without moving the mouse.
- *Drag* means to position the mouse pointer, press the left mouse button and hold it down while you move the mouse, then release the button.

These are the most common mouse pointers used by $SW$:

| Pointer | Use |
| --- | --- |
| I | The I-beam pointer is active when the mouse pointer is over text. |
| ▶ | The arrow pointer is active when the mouse pointer is over mathematics; when menus and dialog boxes are open; and when the pointer moves to the toolbars, the Status bar, and the scroll bars. |
| ✂ | The scissors pointer is active when you drag selected text or mathematics. $SW$ cuts the selection from the current position and inserts it at the location of the pointer when you release the mouse button. |
| ▤ | The copy pointer is active when you press CTRL and drag selected text or mathematics. $SW$ copies the selection from the current position to the location of the pointer when you release the mouse button. |
| 🖐 | The hand pointer appears when you pan graphics. |
| 🍁 | The maple leaf pointer appears when Maple is performing a computation. |

### Keyboard Conventions

We also use standard Windows conventions to give keyboard instructions.

- The key names we use in the manual match the names shown on most keyboards. They appear like this: CTRL, ESC, SHIFT.

- A comma (,) between two key names indicates that you must press the two keys sequentially. For example, ALT, F, O means that you press the ALT key and release it, press F and release it, and press O and release it.

- A plus sign (+) between two key names indicates that you must press both keys at the same time. For example, SHIFT+END means that you press the SHIFT key and hold it down while you press the END key, then release both keys.

- The notation CTRL + **word** means that you must hold down the CTRL key, type the word that appears in bold type after the +, then release the CTRL key. Note that if a letter appears capitalized, you should type that letter as a capital.

# Technical Support

If you can't find the answer to your questions in the documentation or the online Help, contact ITP Technology Services (for users of the Student Edition) or TCI Software Research (for users of the Professional Edition). We urge you to submit questions by electronic mail whenever possible in case our technical staff needs to obtain your file to diagnose and solve the problem.

If you contact us by electronic mail or fax, please provide complete information about the problem you're trying to solve. Our technical staff must be able to reproduce the problem exactly from your instructions. If you contact us by telephone, you should be sitting at your computer with *SW* running and this book close at hand.

Please provide the following information when you contact Technical Support:

- Your *SW* serial number.
- The version number of Windows and the version number of *SW* you're using.
- The type of hardware you're using, including network hardware.
- A description of what happened and what you were doing when the problem occurred.
- The exact wording of any messages that appeared on your computer screen.

▶ **If you have questions about the Student Edition, installing any version of SW, or the basic operation of the software**

- Contact ITP Technology Services by electronic mail, fax, or telephone between 6 a.m. and 5 p.m. Pacific Time:

    **Internet electronic mail address: support@brookscole.com**
    **Fax number: (408) 373-0351**
    **Telephone number: (800) 327-0325**

▶ **If you have questions about the Professional Edition**

- Contact Technical Support at TCI by electronic mail, fax, or telephone:

    **Internet electronic mail address: support@tcisoft.com**
    **Fax number: (505) 522-0116**
    **Telephone number: (505) 522-0352**

## Additional Information from TCI

TCI makes new information available on a regular basis to publicize new developments and answer user requests. When you return your user registration card to TCI, your name is placed on our mailing list so you can receive our newsletter, which contains interesting articles and technical tips. TCI also publishes an electronic newsletter, "TCI News," which contains new product information, answers frequently asked questions, and outlines innovative problem solutions. Visit our home page on the World Wide Web to learn more about TCI and its products.

▶ **To subscribe to TCI News**

- Send a subscription request by electronic mail to **subscribe@tcisoft.com** and include your product serial number.

TCI has established **tcixchange**, an unmoderated electronic mail list, so our customers can share information, discuss common problems, and contribute technical tips and solutions. In addition, if you contact TCI Technical Support by electronic mail, you receive a reply that includes information about obtaining current technical documents by electronic mail.

▶ **To subscribe to tcixchange**

1. Address your electronic mail to **majordomo@list.tcisoft.com**

2. In the body of the message, enter **subscribe tcixchange**

▶ **To cancel your tcixchange subscription**

1. Address your electronic mail to **majordomo@list.tcisoft.com**

2. In the body of the message, enter **unsubscribe tcixchange**

▶ **To visit the TCI Software Research home page**

- Access our home page at **http://www.tcisoft.com/tcisoft.html**

# 1 Understanding the Workplace

The *SW* main window is your workplace. This chapter describes the workplace and explains how to use the commands available through the menus, buttons, popups, dialog boxes, and tab sheets. To display the main window, start the program.

▶ **To start SW**

- Double-click the *SW* icon in the program group.

   –or–

- Select the icon and then press ENTER.

## Using the Main Window

The first thing you see when you start the program is the main window, which has a Menu bar at the top, a Status bar at the bottom, and a series of toolbars around the entry area. Point the mouse at each toolbar button for a few seconds to display a tooltip that identifies the button.

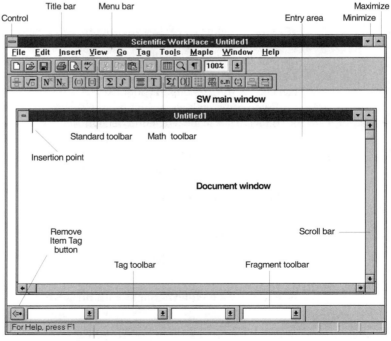

The blinking vertical bar in the entry area is the *insertion point*, which indicates where *SW* will insert the next character you type or where the next command you choose will take effect. The *SW* commands are available through the menus, buttons, and lists shown on the main window, and also from the keyboard. Some *SW* commands have associated dialog boxes from which you can choose additional options.

## Using the Menus

| <u>F</u>ile | <u>E</u>dit | <u>I</u>nsert | <u>V</u>iew | <u>G</u>o | <u>T</u>ag | Too<u>l</u>s | <u>M</u>aple | <u>W</u>indow | <u>H</u>elp |
|---|---|---|---|---|---|---|---|---|---|

The Menu bar lists these menus: File, Edit, Insert, View, Go, Tag, Tools, Maple, Window, and Help. If you use Mathematica for computations (see Chapter 13 "Customizing SW" for more information), the Mathematica menu appears in place of the Maple menu. The computational menus don't appear if you have *Scientific Word*.

When you choose one of these menus, *SW* displays a pull-down menu listing the available commands. You use the commands on the File, Edit, Insert, Tag, and Tools menus to enter your content or define its appearance and the commands on the Go menu to move from place to place. You use the commands on the computation menu to perform mathematical computations. Refer to *Doing Mathematics with Scientific WorkPlace* for complete information about using the Maple menu. You use the commands in the Window and View menus to open and arrange more than one window in your workplace and to arrange the main window and the working view the way you want them. You use the commands on the Help menu to access the online Help feature and online documents.

Any command that appears in the pull-down menus followed by an arrow has a list of associated commands. When you choose a command followed by an arrow, *SW* displays the associated commands so you can choose from among them.

Any command that appears followed by dots has an associated dialog box containing additional options. When you choose the command, *SW* displays the dialog box so you can choose from among the options.

Some commands are available only in certain document environments. For example, if you have not made a deletion, the Undo Deletion command on the Edit menu is not available. Any command that appears dimmed is not currently available.

As with any Windows application, you can make selections from the menus and dialog boxes with the mouse or with the keyboard.

▶ **To choose a menu command with the mouse**

1. Point to a menu name and click the left mouse button.

2. Point to a menu command and click the left mouse button.

▶ **To choose a menu command with the keyboard**

1. Press the ALT key to activate the menu bar.

2. Press the underlined letter for the menu you want.

3. Press the underlined letter for the command you want.

For example, to save a file, press ALT and then press F to open the File menu, then press S to choose Save.

### ▶ To cancel a menu

- Click any point that is off the menu.

  –or–

- Press ESC twice.
  The menu vanishes when you press ESC the first time. It remains active, however, until you press ESC the second time.

## Using the Dialog Boxes

A dialog box is a special window from which you can choose additional options for certain commands. Dialog boxes contain various kinds of controls—option and command buttons, lists, check boxes, and fields for entering text and mathematics.

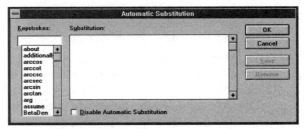

A darkened control indicates that the option is currently active. A dimmed control indicates that it is not currently available. In some dialog boxes, an insertion point appears as a blinking vertical line, as it does in the main window. In others, a small dotted box indicates where your next action will take place.

Some dialog boxes group related commands on a series of *tab sheets*. The controls on tab sheets behave just like those in dialog boxes, with one important difference: *When you choose to accept or discard the new selections on one tab sheet in a dialog box, you accept or discard any new selections on all tab sheets in the dialog box.*

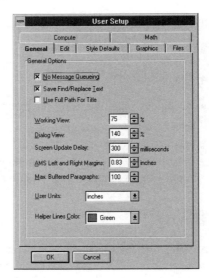

For example, if you make the settings shown above in the General tab sheet of the User Setup dialog box, then move to the Edit tab sheet and press Cancel, you will discard the changes you made on the General tab sheet.

Most dialog boxes have OK and Cancel buttons. You use them to indicate that you have finished making selections in the dialog box. In a dialog box with tab sheets, you use the OK and Cancel buttons to accept or discard all the selections you have made, not only in the active tab sheet but also in any other tab sheets in the dialog box. When the changes you make on a tab sheet can't be undone without reopening the dialog box, the Cancel button changes to Close.

- Choose the OK button to have *SW* accept and act on any choices you've made anywhere in the dialog box and return to the document.

- Choose the Cancel button to discard any choices you've made anywhere in the dialog box and return to the document.

▶ **To select an item from a list in a dialog box**

1. Click the arrow to the right of the list box.

2. Select the item you want from the list that appears.

▶ **To choose OK**

- Click OK.

  –or–

- Choose the OK button, then press ENTER.

► **To choose** Cancel

- Click Cancel.

  –or–

- Press ESC.

  –or–

- Choose the Cancel button, then press ENTER.

## Using the Toolbars

When you start *SW*, a series of toolbars appears in the window. The buttons on the toolbars are identical in function to many of the commands on the File, Edit, Insert, View, Go, and Maple menus. Appendix A "Toolbar Buttons and Menu Commands" lists the equivalent menu commands for each button.

At times you may want to hide some of the toolbars so you can see more of your document in the window. The Toolbars command on the View menu opens a dialog box containing toggle commands for hiding and showing the toolbars. To learn more about arranging the main window the way you want, see Chapter 13 "Customizing SW."

► **To choose a toolbar button**

- Click it with the mouse.

  Those buttons without dots have an immediate effect; those with dots open an associated dialog box containing more options.

► **To select an item from a list**

1. Click the arrow to the right of the list box.

2. Select the item you want from the list that appears.

► **To show or hide a toolbar**

1. From the View menu, choose Toolbars (ALT+V, T).

2. In the Toolbars dialog box, check the box next to each toolbar that you want to display and uncheck the box next to each toolbar that you want to hide.

3. Choose OK.

### The Standard Toolbar

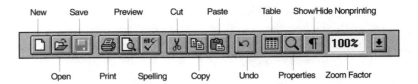

The buttons on the Standard toolbar invoke common file, document, and editing operations and change the screen appearance of the document in the active window.

### The Math Toolbar

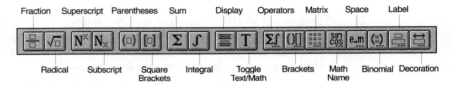

The buttons on the Math toolbar insert mathematical objects directly into your document or open dialog boxes for mathematical objects that can't be inserted into your document without more information. The toolbar also includes a special tool, the Math/Text toggle.

### The Symbol Toolbar

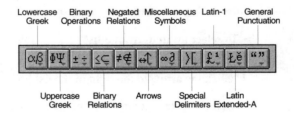

Each button on the Symbol toolbar has a popup panel containing mathematical, Greek, or Unicode symbols and characters that you can insert in your document. Appendix A "Toolbar Buttons and Menu Commands" shows the symbols found in each panel.

▶ **To insert a symbol from a panel into your document**

1. Click the button for the panel containing the symbol you want.

2. Click the symbol in the panel.

## The Common Symbols Toolbar

The Common Symbols toolbar contains buttons for many frequently used mathematical symbols.

▶ **To enter a symbol from the Common Symbols toolbar into your document**

• Click the symbol you want.

## The Compute Toolbar

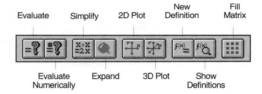

In *Scientific WorkPlace*, the main window has a Compute toolbar. The buttons invoke the most common computational operations. Refer to *Doing Mathematics with Scientific WorkPlace* for information about performing computations.

▶ **To invoke a computational command**

1. Select the expression you want to compute, or place the insertion point to its right.

2. Click the button for the operation you want.

## The Navigate Toolbar

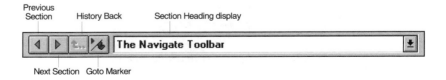

The Navigate toolbar helps you move quickly through your document by jumping between section headings and from marker to marker. The wide box displays the heading of the section currently containing the insertion point.

## The Field Toolbar

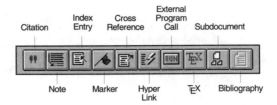

The Field toolbar contains buttons for enhancing the structure of your document with internal references such as citations, notes, cross-references, and index entries; with hypertext links to other parts of the document; with links to subdocuments; and with external program calls. Each button on the toolbar opens a dialog box.

## The Tag Toolbar

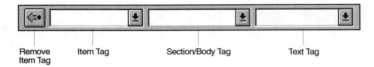

A *tag* is a field that adds structure or content to a paragraph or text selection. The Tag toolbar contains popup lists for the *SW* tags:

- *Item tags* create bibliography items, different kinds of lists, and theorem and theorem-like statements.

- *Section tags* create section headings and provide the structure of your document.

- *Body tags* apply structure to the text in the body of your document.

- *Text tags* emphasize or differentiate words or phrases within a paragraph.

The three tag boxes also name the tags in effect at the insertion point. The **Remove Item Tag** button at the left end of the Tag toolbar removes the most recently inserted item tag in the paragraph containing the insertion point. The tags that appear in each popup list depend on the typesetting style of your document. Each tag is also associated with a predefined print appearance and a predefined screen appearance, which also depend on the typesetting style. You can also apply tags with the function keys and by using the **Apply** command on the **Tag** menu.

▶ **To select a tag from a popup list**

1. Click the popup box to display the list.

2. Click the tag you want.

## The Fragments Toolbar

A *fragment* is a portion of a paragraph saved in a separate file for later recall. *SW* provides you with many predefined fragments. In addition, you can create your own fragments by saving portions of text or mathematics. The names of all available fragments appear in the popup list on the Fragments toolbar. The fragments in the list are also available from the Import Fragment dialog box. See Chapter 6 "Using Special Features" for more information on fragments.

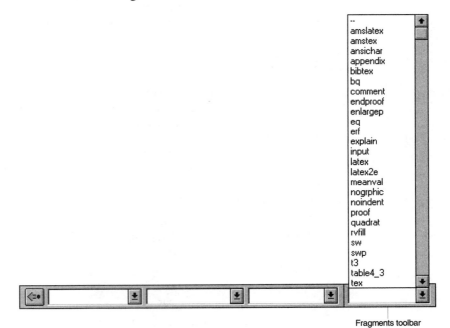

Fragments toolbar

▶ **To enter a fragment**

1. Click the Fragments popup box to display the popup list.

2. Click the fragment you want.

   –or–

1. From the File menu, choose Import Fragment (ALT+F, R).

2. Select the fragment you want.

3. Choose OK.

   –or–

● Press CTRL+*name,* where *name* is the name of the fragment you want.

## Moving Around in the Workplace

Inside *SW* documents, actions take place at the insertion point, unless you have marked, or *selected,* a part of your document by highlighting it. The insertion point is where the program will insert the next character you type or execute the next command you choose.

When you're entering content, *SW* makes room for the new information at the insertion point, automatically scrolling through the document and displaying new portions of the document as necessary to keep the insertion point visible in the main window. When you're adding to existing material, you may need to scroll through the document, moving the insertion point from place to place in your document.

## Scrolling

You can scroll rapidly through the length of your document with the keyboard or the mouse. Because *SW* automatically fits your document to the width of the window, you need to scroll horizontally only when your document contains an unbreakable object that is wider than the window.

To scroll forward and backward through your document using the keyboard, you press certain keys or combinations of keys. To scroll with the mouse, you drag the shaded box on the vertical *scroll bar* of the document window. The box on the bar indicates the approximate location in the document of the text displayed in the window.

| To | With the mouse | With the keyboard |
|---|---|---|
| Scroll up | Drag the shaded scroll box up on the scroll bar. | Press UP ARROW repeatedly. |
| Scroll down | Drag the shaded scroll box down on the scroll bar. | Press DOWN ARROW repeatedly. |
| Scroll to the top of a document | Drag the shaded scroll box to the top of the scroll bar. | Press CTRL+HOME. |
| Scroll to the bottom of a document | Drag the shaded scroll box to the bottom of the scroll bar. | Press CTRL+END. |
| Scroll one screen up | Click the scroll bar above the shaded scroll box. | Press PAGE UP. |
| Scroll one screen down | Click the scroll bar below the shaded scroll box. | Press PAGE DOWN. |
| Scroll one line up | Click the arrow at the top of the scroll bar. | Press UP ARROW until the line appears. |
| Scroll one line down | Click the arrow at the bottom of the scroll bar. | Press DOWN ARROW until the line appears. |

## Moving the Insertion Point

When you scroll to a different part of your document, you may also need to move the insertion point. To move the insertion point in your document, you can use navigation commands or you can move with the mouse, the keyboard, or, if NUM LOCK is off, the directional arrows on the numeric keypad. You can insert hypertext jumps so you can move rapidly through the text. See Chapter 6 "Using Special Features" for more information.

### Navigating

The navigation commands on the Navigate toolbar and the **Go** menu speed moving from place to place in your document. You can move quickly to a particular section, to the beginning of a particular paragraph, or to the beginning of the paragraph that contains a particular marker.

▶ **To move to a particular section**

1. On the Navigate toolbar, click the Section Heading popup list.

   The Section Heading box lists all the headings in the document and highlights the heading of the section that contains the insertion point.

2. On the list, click the section heading you want.

   The program moves to the section and places the insertion point at the beginning of the heading.

▶ **To move to the previous section**

- On the Navigate toolbar, click the Previous Section button  .
  −or−
- From the **Go** menu, choose **Previous Section** (ALT+G, P).

▶ **To move to the next section**

- On the Navigate toolbar, click the Next Section button ▷ .
  −or−
- From the **Go** menu, choose **Next Section** (ALT+G, N).

▶ **To move to the source of the previous jump**

- On the Navigate toolbar, click the History Back button ⬆ .
  −or−
- From the **Go** menu, choose **History Back** (ALT+G, B).

▶ **To move to the paragraph or object containing a particular marker**

1. On the Navigate toolbar, click the Goto Marker button .

   –or–

   From the Go menu, choose Goto Marker.

2. From the list in the dialog box, select the marker to which you want to move the insertion point.

3. Choose OK.

▶ **To move to a particular paragraph**

1. From the Go menu, choose Goto Paragraph.

   The Goto Paragraph dialog box displays the number of the paragraph that contains the insertion point.

2. In the dialog box, specify the number of the paragraph to which you want to move the insertion point.

3. Choose OK.

## Moving with the Mouse and the Keyboard

▶ **To use the mouse to move the insertion point in the main window**

- Point the mouse where you want the insertion point to be and click.

▶ **To use the keyboard to move the insertion point**

- Use the commands in the following table.

| To move the insertion point | Press |
|---|---|
| To the left | LEFT ARROW |
| To the right | RIGHT ARROW (and SPACEBAR in math) |
| To the start of the next word or object | CTRL+RIGHT ARROW |
| To the start of the previous or current word or object | CTRL+LEFT ARROW |
| Up | UP ARROW |
| Down | DOWN ARROW |
| To the start of the line | HOME |
| To the end of the line | END |
| To the next field inside a template | TAB or INSERT |
| To the previous field inside a template | SHIFT+TAB |
| To the outside of a template | RIGHT or LEFT ARROW or SPACEBAR |
| Next screen | PAGE DOWN |
| Previous screen | PAGE UP |
| To the beginning of the document | CTRL+HOME |
| To the end of the document | CTRL+END |

## Moving Around in Dialog Boxes

You can move around and make selections in a dialog box with the mouse or, using the keys in the following table, with the keyboard.

| To | Press |
|---|---|
| Move the insertion point to the next control | TAB |
| Move the insertion point to the next option | ARROW |
| Move the insertion point to the previous control | SHIFT+TAB |
| Move the insertion point directly | ALT+underlined letter in the name |
| Display a drop-down list | ALT+underlined letter in the name, F4 |
| Select from a list | UP ARROW or DOWN ARROW, then ALT+DOWN ARROW to close the list |
| Close a list | ALT+DOWN ARROW |
| Adjust a number up or down | UP ARROW or DOWN ARROW |
| Check or uncheck a box | SPACEBAR |
| See the start of a range of numbers | CTRL+HOME |
| See the end of a range of numbers | CTRL+END |
| Move to a different tab sheet | TAB to highlight the current tab sheet name, then RIGHT ARROW or LEFT ARROW |

## Making Selections

*SW* executes the commands you give it at the insertion point unless you have selected part of your document. If you have made a selection, the next action you take will affect the entire selection. You can select part of a document with the mouse or with the directional arrows in combination with other keys on the keyboard.

▶ **To select part of a document with the mouse**

1. Point to where you want the selection to begin.

2. Hold down the left mouse button.

3. Drag the mouse pointer to where you want the selection to end.

4. Release the mouse button.

▶ **To select a large part of a document with the mouse**

1. Point to where you want the selection to begin and click the mouse.

2. Press and hold SHIFT.

3. Move the mouse pointer to where you want the selection to end.

4. Click the mouse.

Note that you may need to scroll to move the mouse pointer to where you want the selection to end.

▶ **To make a selection with the keyboard**

- Use the commands in the following table.

| To select | Press |
|---|---|
| The character to the right | SHIFT +RIGHT ARROW |
| The character to the left | SHIFT+LEFT ARROW |
| The word to the right | CTRL+SHIFT+RIGHT ARROW |
| The word to the left | CTRL+SHIFT+LEFT ARROW |
| To the end of the line | SHIFT+END |
| To the start of the line | SHIFT+HOME |
| One line up | SHIFT+UP ARROW |
| One line down | SHIFT+DOWN ARROW |
| To the end of the document | CTRL+SHIFT+END |
| To the start of the document | CTRL+SHIFT+HOME |

# Getting Help

While you're working in *SW*, you can get information quickly from the commands on the online Help menu. The commands provide access to online Help and to the online manuals and tutorials.

## Using Online Help

You can use online Help to get information about the installed version of *SW*, information about all *SW* commands and operations except those involving Maple computations, and instructions for using online Help. You can keep the Help information in view while you work on your document, and you can flag and annotate frequently used Help topics.

▶ **To open online Help**

- Press F1 to go directly to the Help contents.
  –or–
- From the Help menu, choose
    - **Contents** (ALT+H, C) to see a list of the Help topics.
    - **Keyboard Accelerators** (ALT+H, K) to get information about using the keyboard to enter text and mathematics quickly.
    - **Search for Help on** (ALT+H, S) to open the index of all Help topics.
    - **How to Use Help** (ALT+H, H) to get information about online Help itself.
    - **About SW** (ALT+H, A) to get information about *SW*.

▶ **To close the Help window**

- From the File menu in Help, choose **Exit** (ALT+F, X).

### Understanding the Help Window

The Help window has a menu bar and several buttons you can use to move around in online Help. When a button or menu command is not active, its name is dimmed.

The online information often contains underlined words and phrases called *jump terms*. If you click a jump term, the corresponding Help information is displayed.

Some terms in the online information have dotted underlines to indicate they are *glossary terms*. If you click a glossary term, online Help displays the definition of the term.

### Finding the Information You Need

You can move directly to the information for any jump term that appears in the Help window, or you can search the Help topics for information on any topic you choose. You can move from one Help topic to another and then retrace your steps.

▶ **To move directly to a topic**

- Click the jump term for the topic.

▶ **To search for a topic**

1. Choose Search for Help on (S).

2. Type a word or phrase describing the information you want

   –or–

   Select a topic by scrolling through the list of topics.

3. Choose Display (ALT+D).

   The program displays a list of those topics related to the one you chose.

4. Select the topic you want.

5. Choose Display (ALT+D).

   Help displays the online information you want.

▶ **To retrace your steps**

• Choose Back (B).

## Using the Online Manuals and Tutorials

Through the Help menu you can open the documentation provided with *SW* and an extensive set of tutorial exercises for learning how to work with documents and mathematics.

▶ **To open the online manuals and tutorials**

• From the Help menu, choose

   - Getting Started to open an online copy of *Getting Started with Scientific Work-Place and Scientific Word* (ALT+H, G).
   - Creating Documents to open *Creating Documents with Scientific WorkPlace and Scientific Word* (ALT+H, R).
   - Doing Mathematics to open *Doing Mathematics with Scientific WorkPlace* (ALT+H, D).
   - Tutorial to open the tutorial exercises (ALT+H, T).

   The program opens the online information you want in a new window.  You can scroll through these documents as you would through any open document; see "Moving Around in the Workplace" earlier in this chapter.  The online manuals and tutorials also contain *hypertext links* to other points in the document. Hypertext links dramatically speed jumping from topic to topic within a document. The links appear on the screen in green. See Chapter 6 "Using Special Features" for more information.

▶ **To use hypertext links to move through a document**

- Hold down the CTRL key and click the mouse in a hypertext link.
  –or–
- Click in a hypertext link and then from the Tools menu, choose Action (ALT+L, A).
  The program jumps to the linked topic and places the insertion point at the beginning of the paragraph.

▶ **To move back to the source of the hypertext link**

- On the Navigate toolbar, click the History Back button.
  –or–
- From the Go menu, choose History Back (ALT+G, B).
  The program jumps back to the source of the link and places the insertion point at the beginning of the paragraph.

# 2 Opening and Closing Documents

To work in *SW*, you must open a document. You can then enter and edit text and mathematics in the document, perform computations (in *Scientific WorkPlace*), and preview and print the results. You must save your document to preserve the work you do. Saving a document stores it, in its current state, as a file on your hard drive or on a floppy disk. Remember to save your documents regularly.

## Opening Documents

Starting *SW* automatically opens a document window containing an empty start-up document temporarily named Untitled1. You can start working immediately in the start-up document, open a new document with a different style, or open a document that already exists. With *SW* you can open several documents at once or several views of the same document.

### Working in the Start-up Document

The start-up document has a default *document type* (such as book, article, examination, or memo) and a default *document style* that determines how the document will appear in print. You can preview the style's typeset appearance from the File menu.

▶ **To preview the appearance of the default style**

1. From the File menu, choose Style (ALT+F, Y).

2. Choose Select Style (ALT+S).

3. Choose Preview Sample (ALT+P).

    Scroll through the preview document to see how it looks in print.

4. To leave the preview, choose Exit from the File menu (ALT+F, X).

5. Choose Cancel to leave the Predefined Styles dialog box.

6. Choose Cancel to leave the LaTeX Styles for Print and Preview dialog box.

---

**Note**   You can specify a different default style for start-up documents from the User Setup dialog box. See Chapter 13 "Customizing SW" for instructions.

---

If you want to start a document with the default type and style, you can begin entering content right away. Otherwise, open a new document and select the style you want.

## Opening a New Document

To start a document with a different type and style, open a new document.

---

**Note**    Choose the new document type with care. All document types do not have the same elements, and converting to a different document type at a later time can cause unpredictable results. See Chapter 10 "Applying Styles to Documents" for information about changing styles.

---

▶ **To open a new document**

1. On the Standard toolbar, click the New button 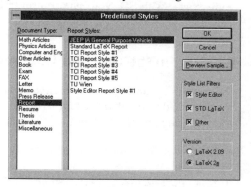.

   –or–

   From the File menu, choose New (ALT+F, N).

   The program displays the Predefined Styles dialog box:

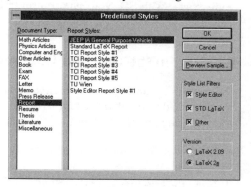

2. Choose the document type you want.

   The program displays the available typesetting styles for the type you choose.

3. Choose the typesetting style you want.

4. To see how the style appears in print, choose Preview Sample (ALT+P).

   Scroll through the sample document. When you're ready to leave the preview, choose Exit from the File menu to return to the Predefined Styles dialog box.

5. Choose OK.

   The program loads the *shell document* associated with the chosen style and temporarily names the file as the next sequential Untitled file. You give the document the name you want when you save it. If you have just started *SW* and if the start-up document has not been modified, the program discards the start-up document.

Some shell documents contain predefined information; depending on the style, that information may appear in the document window. You can use any *SW* document as a shell.

## Opening an Existing Document

You can open any existing *SW* file with the **Open** command on the **File** menu or the Open button on the Standard toolbar. Also, you can open the files you have worked on most recently by choosing a file name from the short list of documents that appears at the bottom of the **File** menu. If your document is listed on the menu, this is the fastest way to open it.

► **To open a recently opened document**

1. Choose the **File** menu (ALT+F).

   The program lists the most recently opened documents at the bottom of the menu.

2. Click the name of the document you want.

   –or–

   Type the underlined number next to the name of the document you want.

   The program opens the document you selected and, if you have just started *SW*, discards the start-up document if it has not been modified.

► **To open an existing document**

1. On the Standard toolbar, click the Open button  .

   –or–

   From the **File** menu, choose **Open** (ALT+F, O).

2. Select the document you want from the **File Name** list, or type the name of the document in the **File Name** box.

   If you do not see the name of the document you want, you may need to scroll through the list of documents. If the document does not appear in the list, you may need to select a different directory from the **Directories** box, select a different drive from the **Drives** box, or select a different type of file from the **List Files of Type** box.

   If you're working on a network and you need to connect to a different network directory, choose **Network**.

3. Choose **OK**.

   The program opens the document you selected and, if you have just started *SW*, discards the start-up document if it has not been modified.

## Opening a Document with a Different Format

With $SW$ you can open existing documents in other file formats, including $T^3$, LaTeX, ASCII, ANSI, and Rich Text Format (RTF). Documents in some file formats must be manually converted to TeX before you open them; others are converted automatically. Regardless of the original file format of a document, $SW$ will store it as a LaTeX file when you save the document.

---

**Note**   You can use the ASCII import feature of other word processors to import LaTeX files. When you do this, mathematics appears in LaTeX code form. For example, $\alpha^2 + \beta^2$ appears as $\alpha^2+\beta^2$.

---

▶ **To open a document with a different format**

1. On the Standard toolbar, click the Open button  .

   –or–

   From the File menu, choose **Open** (ALT+F, O).

2. In the **List Files of Type** box, select the format of the document you want to open.

3. In the **Drives** and **Directories** boxes, select the location of the document.

4. In the **File Name** box, type or select the name of the document.

5. Choose **OK**.

## Opening T³ Files

$SW$ has a special filter for opening a document created with the $T^3$ Scientific Word Processing System, TCI's DOS product. The document must first be converted to TeX using $T^3$'s file conversion utility. This utility, called T3toTeX, is found in $T^3$ version 2.3.

▶ **To transfer a T³ document to DOS**

1. From the $T^3$ **Main** menu, accept **Document**.

2. Select the document you want to convert.

3. From the **Operations** menu, accept **DOS Transfer**.

   This opens the **Transfer Document** form. This example assumes you create a DOS file called `my.t3` in the directory `c:\swp25\convert`. The latter directory contains the conversion utilities that are installed when you install $SW$. If the output file name does not include the complete path, the file is placed in the current directory. Use the control file `convert.cnv`. Highlight all the parts of the document in the bottom field of the form.

4. Fill in the form and then accept it.

When the transfer process is complete, $T^3$ returns to the **Main** menu.

5. Exit $T^3$.

6. At the DOS prompt in the `c:\swp25\convert` directory, type

**t3totex    my**

to finish converting the $T^3$ file to TEX. The result of the conversion is a Plain TEX file. *SW* has a special filter that can import this special, restricted form of Plain TEX.

▶ **To open the Plain TeX file in SW**

1. Start *SW*.

2. On the Standard toolbar, click the Open button .

–or–

From the **File** menu, choose **Open** (ALT+F, O).

3. Click the down arrow next to the **List Files of Type** box.

4. Choose **T3 to TeX**.

5. Use the other list boxes in the dialog box to change to the `c:\swp25\convert` directory and select `my.tex`.

6. Choose **OK**.

7. When you've finished making changes, save the file.

If the source $T^3$ document contains built-up mathematical expressions that cannot be converted to TEX, the converter outputs a column of characters using the macro `\tcol`. When encountered by *SW*, `\tcol`s are contained in TEX fields. These are easy to see in *SW* and can be replaced by the correct construct using *SW*'s editing features. We suggest you have a $T^3$ printout of your document at hand when you perform this editing, as the most effective approach is to delete the `\tcol`s and replace them. There is no effective way to use the text in the `\tcol` fields.

Some $T^3$ documents can produce strange results when converted to TEX and opened in *SW*. For example, $T^3$ allows the placement of a superscript character in the subscript position, producing the following in TEX:

$$x_{^2}$$

Although this prints correctly in $T^3$, it does not look good in TEX and you must change it to a simple subscript in *SW*. More detailed information about what the conversion does can be found in the notes for the $T^3$ Conversion Programs included with the $T^3$ documentation.

### Opening Existing LaTeX Files

Some existing LaTeX documents can be opened directly by *SW*. Some constructs, however, need modification with an ASCII editor before they can be opened. In general, if LaTeX has a construct that differs from Plain TeX (`array` vs `matrix`, for example), the LaTeX construct should be used. Statements of the form `\newcommand` and `\def` usually cause problems. As a defensive measure, *SW* expands these statements, which is generally not what you want. Refer to the document `techref.tex` in the `extras` directory for more information about this issue. If a document is successfully read by *SW*, modified, and saved, the command `\input{tcilatex}` is added to the preamble.

### Opening ASCII and ANSI Files

*SW* has ASCII and ANSI filters for reading arbitrary files. Characters that are not valid in LaTeX are removed, and characters that need special handling ({,},$,#,@,%,&,\,|,<,>, ~) are dealt with correctly. The ANSI filter converts the upper ANSI characters to the appropriate LaTeX characters.

▶ **To open an ASCII or ANSI file in SW**

1. On the Standard toolbar, click the Open button .

   –or–

   From the File menu, choose **Open** (ALT+F, O).

2. Click the down arrow at the right end of the **List Files of Type** box.

3. Click **ASCII** or **ANSI**.

   Choose ANSI if the file you want to read contains characters from the upper ANSI set that you want to preserve.

4. Use the other controls in the dialog box to change to the directory containing the file and to select the file.

5. Choose **OK**.

6. When you've made all necessary changes, save the file.

### Opening Rich Text Format (RTF) Files

Many word processing systems can export files in the Microsoft RTF format. *SW* includes a filter for reading RTF files. The RTF filter converts mathematics created with the Word equation editor.

▶ **To read a plain text RTF file into SW**

1. On the Standard toolbar, click the Open button .

   –or–

From the **File** menu, choose **Open** (ALT+F, O).

2. Click the down arrow at the right end of the **List Files of Type** box.

3. Click **RTF**.

4. Use the other list boxes in the dialog box to change to the directory containing the file and to select the file.

5. Choose **OK**.

6. When you've made all necessary changes, save the file.

## Opening Several Documents at Once

You can open several different documents or several views of the same document at the same time using different *SW* windows. When you have several windows open, the title bar of the active window is displayed in the same color as the *SW* title bar. You can move easily from window to window, and you can copy and move information from document to document. If you open multiple views of the same document, the changes you make in one view are recorded in all views. Also, you can change the way you view the document in each window and the way you arrange the windows in *SW*. You can minimize any document window so that it appears as an icon in the *SW* main window. See Chapter 13 "Customizing SW" for details.

▶ **To open an existing document**

1. On the Standard toolbar, click the Open button  .

   –or–

   From the **File** menu, choose **Open** (ALT+F, O).

2. In the **Open** dialog box, specify the file you want to open.

3. Choose **OK**.

   *SW* opens the document into a new window and places the insertion point at the beginning of the document.

▶ **To open a new document**

1. On the Standard toolbar, click the New button  .

   –or–

   From the **File** menu, choose **New** (ALT+F, N).

2. From the **Predefined Styles** dialog box, choose the style you want.

3. Choose **OK**.

*SW* opens a new window and places the insertion point in it.

▶ **To open another view of the active document**

- From the **Window** menu, choose **New Window** (ALT+W, N).
  *SW* opens the active document into another window and places the insertion point at the beginning of the document.

▶ **To move from window to window**

- From the list at the bottom of the **Window** menu, choose the number next to the window you want.
  –or–
- Click the mouse on the title bar of the document window you want.

▶ **To maximize a document window**

- Click the maximize button at the top right of the document window.
  –or–
- Double-click the title bar of the document window.
  –or–
- From the control box at the top left of the document window, choose **Maximize** (ALT+SPACE, RIGHT ARROW, X).
  When a document window is maximized, the minimize button and the control box for the document appear at opposite ends of the Menu bar.

▶ **To minimize a document window**

- Click the minimize button at the top right of the document window.
  –or–
- From the control box at the top left of the document window, choose **Minimize** (ALT+SPACE, RIGHT ARROW, N).

# Saving Documents

To preserve your latest work, save your document regularly. Saving your document regularly means that you lose very little work if you experience problems. We suggest that you save your document every 10 to 20 minutes and also before you print, before you make a large change (such as a global replace), and after you enter any work that would be difficult to redo. To ensure that your work is saved routinely, you can have *SW* automatically save your documents as frequently as you want. See "Saving Automatically at Regular Intervals" later in this chapter. When you save a document, it remains open so you can keep working.

## Saving an Existing Document

When you choose the **Save** command, *SW* saves the open document with the same name and in the same directory you used the last time you saved the document.

▶ **To save an existing document**

- On the Standard toolbar, click the Save button .
  –or–
- From the **File** menu, choose **Save** (ALT+F, S).

  If you want to save the document with a new name or in another location, save a copy of the document. See "Saving a Copy of an Open Document" later in this chapter.

## Saving a New, Untitled Document

When you save an untitled document, the program automatically opens the **Save As** dialog box so you can name the document. You can assign to your document any DOS file name that is acceptable for your installation. For example, if you have installed *SW* on a Win32s system, the file name can be from one to eight characters long followed by an optional period and a one- to three-character extension. An *SW* file name can't be the same as a DOS command name. Otherwise, the name can include any characters except spaces and the following: **\* ? , ; [ ] + = \ / : | < >**. We suggest that you also avoid the use of characters that have special meanings on other platforms. For example, $ and # have special meanings (as do many other nonalphanumeric characters) in certain Unix shells. You can use a period only to separate the file name from the extension.

File extensions distinguish between types of files. *SW* uses `.tex` for documents, `.bak` for backups of documents, and `.shl` for shell documents. See Chapter 12 "Managing Documents" for a complete list of file extensions used by the program. *SW* automatically adds a `.tex` extension to the file name unless you type another extension. You can use a different extension by typing a period and the extension you want, such as `article.ltx`. Note, however, that if you use other extensions, you must include the extension every time you type the file name. If you do not want any extension, type the file name and end it with a period. Regardless of the extension, *SW* saves all documents as LaTeX files.

▶ **To save a new, untitled document**

1. On the Standard toolbar, click the Save button .

   –or–

   From the **File** menu, choose **Save** (ALT+F, S) or **Save As** (ALT+F, A).

   The program displays the **Save As** dialog box.

2. In the **File Name** box, type a name for your document.

   The program adds a `.tex` extension to the name unless you type another extension.

3. If you want to save the document on another drive or in another directory, do one of the following:

- From the **Drives** and **Directories** boxes, select the drive and directory you want.
  –or–
- Type the complete path in the **File Name** box.
  Example   `a:\docs\chap1`

If you're working on a network, you may need to save your file to the network file system; choose **Network**.

4. If you want to save the document as a read-only file, so that the document can be opened and read but not changed, check **Read Only**.

5. Choose **OK**.

If you type a file name that already exists in the chosen directory, the program asks whether you want to replace the existing document with the new document.

- Choose **No** to preserve the old document and then type another file name.
  –or–
- Choose **Yes** to replace the old document with the open document.
  If the old document is a read-only document, *SW* asks if you want to replace it. If you want to preserve the old document, choose **No** and enter a different file name.

## Saving a Copy of an Open Document

If you want to create a copy of the document with a new name or save the document in another location, use the **Save As** command.

▶ **To save a copy of the open document under a new name or in a new location**

1. From the **File** menu, choose **Save As** (ALT+F, A).

2. To save the document under another name, type a new name for the document in the **File Name** box.

   The previously saved version of the document is available under the old name.

3. To save the document on a different drive or in a different directory, do one of the following:

- From the **Drives** and **Directories** boxes, select the drive and directory you want.
  –or–
- Type the complete path for the document file in the **File Name** box—for example, `a:\revision\revchap1`.

If you're working on a network and you want to save a copy of the document to the network file system, choose **Network** and then specify the destination network drive.

4. Choose **OK**.

If you type a file name that already exists in the chosen directory, the program asks whether you want to replace the existing document with the new document.

- Choose **No** to preserve the old document and then type another file name.
  –or–
- Choose **Yes** to replace the old document with the open document
  If the old document is a read-only document, *SW* asks if you want to replace it. If you want to preserve the old document, choose **No** and enter a different file name.

## Saving All Open Documents

When you have multiple documents open, you can save them all with a single command.

▶ **To save all open documents**

- From the **File** menu, choose **Save All** (ALT+F, V).
  *SW* saves each open document with the same name and in the same directory used the last time it was saved. If any open documents are new and untitled, the program opens the **Save As** dialog box so you can name the file.

## Using Special Saving Techniques

You can use several special techniques when you save your document. Saving a quick-load form of your document minimizes the time needed to open the document, and creating a read-only document file prevents accidental changes.

### Creating Quick-Load Files

The *SW* Quick-Load feature uses a special form of the document file to open documents more quickly than usual. The feature is especially useful for opening large documents. When a quick-load file is being opened, the paragraph numbers do not appear in the Status bar.

You must save a document for quick loading before you can open it with the Quick-Load feature. Documents saved for quick loading have the same name as the document file but have a file extension of .cdx instead of .tex. You can save your document in both forms. Subdocuments can't be saved with the Quick-Load feature.

---

**Important**    A quick-load file is typically two to three times larger than its original LaTeX document.

---

▶ **To save a document for quick loading**

1. From the File menu, choose Document Info (ALT+F, D).

2. On the General tab sheet, check the Save For Quick Load box.

3. If you don't want to save the .tex form of your document as well as the quick-load form, uncheck Export TeX File On Save.

4. Save the document.

## Creating Read-Only Files to Prevent Unwanted Changes

*SW* provides the Read-Only option to keep you from accidentally making changes to a document that you want to preserve. Many commands are not available in a read-only environment; they appear dimmed on the menus. When you try to save a read-only document that you have modified, the program warns you that the document is protected and gives you the option of overriding the protection or saving the changes in a document with a new name.

▶ **To save a document as read only**

1. From the File menu, choose Save As (ALT+F, A).

2. If you want the read-only version to have a different name or if the document is untitled, type a name for the file.

3. Check Read Only.

4. Choose OK.

▶ **To remove the read-only protection**

1. From the File menu, choose Save As (ALT+F, A).

2. Leave the file name unchanged.

3. Leave the Read Only box unchecked.

4. Choose OK.

5. When the system asks if you want to replace the file,

   - Choose Yes to replace the file.
   - Choose No to retain the read-only file unchanged.

   If you choose to replace the file, the program opens the Overwrite Read Only File dialog box.

6. Choose the way you want to save the file:

- To update the file and save it as a read-only file, check **Read Only File**.
- To update the file and save it as a read/write file, check **Read/Write File**.

7. Choose **OK**.

▶ **To save to a read-only file**

1. On the Standard toolbar, click the Save button .

   –or–

   From the **File** menu, choose **Save** (ALT+F, S).

2. When the system asks if you want to replace the file,

- Choose **Yes** to replace the file.
- Choose **No** to retain the read-only file unchanged.

   If you choose to replace the file, the program opens the **Overwrite Read Only File** dialog box.

3. Check **Read Only File** to update the file and maintain it as a read-only file.

4. Choose **OK**.

## Preserving Documents

You should regularly store copies of your documents on a file server, on floppy disks, or on a tape backup. If you store copies of your documents regularly, you have another copy of your work if a problem occurs with your computer.

In addition to saving documents with the standard saving procedures described earlier in this chapter, you can enhance the safety of your documents by saving documents automatically and creating backup files. These features are available from the **Files** tab sheet in the **User Setup** dialog box. We recommend you use them both.

- The Automatic Saving feature saves the open document regularly at an interval you set. This option regularly saves a complete copy of your document between the times you choose **Save, Save As,** or **Save All.** If a power outage or a system failure occurs, you lose only the changes you have made to the open document since the most recent automatic or manual save.

- The Create Backup Files feature automatically creates a backup copy of the previously saved version every time you save an open document. Each new backup copy replaces the previous one.

## Saving Automatically at Regular Intervals

If you choose the Automatic Saving feature, *SW* saves all open documents at regular intervals, the length of which you choose. *SW* saves each open document in a file that has the same name as the document file and a file extension of .aut. When you open an *SW* document, the program opens the most recent version of the document. If the most recent version has a file extension of .aut, the program asks if you want to open it instead of the .tex version.

---

**Note**   The Automatic saving feature works only with named files. When you create a new document, you must name it before *SW* can save it automatically at the regular intervals you specify.

---

▶ **To save all open documents automatically**

1. From the Tools menu, choose User Setup (ALT+L, U).

2. Choose the Files tab.

3. In the Miscellaneous area, check the box next to Automatic Saving.

4. Type or select how often (in minutes) you want automatic saving to take place.

   You can choose any time interval in the range from 1 to 100 minutes. The initial default is 20 minutes.

5. Choose OK.

   Each time you save a file manually (using the Save button, Save, Save As, or Save All), the automatic saving counter is set to zero. This means that if you always save manually before the automatic saving time interval is up, *SW* never automatically saves your document.

## Creating Backups Automatically

If you choose the Create Backup Files feature, *SW* makes a copy of the previously saved version of the open document each time the document is saved. The backup copy has the same name as the original document but a different file extension. Backup copies of document files have a .bak extension. Each new backup copy for a document file replaces the previous one.

Although creating backup files protects your document while you work, it does not provide long-term safety because the .bak file is saved in the same directory and on the same drive as the original file. Any problems with the disk can damage both the original and the backup. Remember to make copies of your documents regularly on a file server, on floppy disks, or on a tape backup.

▶ **To make a backup copy each time the open document is saved**

1. From the Tools menu, choose User Setup (ALT+L, U).

2. Choose the Files tab.

3. In the Miscellaneous area, check Create Backup [.BAK] Files.

4. Choose OK.

## Restoring a Document

If you cannot open an *SW* document because of damage by a power failure or some other computer problem and Create Backup [.BAK] Files was in effect when the problem occurred, you can open the backup copy of the document. It is the next-to-last version of the original file. The backup document has the same name as your original document but with the extension .bak instead of .tex. For example, if your document was named precious.tex, the backup document is named precious.bak.

If a system problem occurs while you're working in a document, *SW* attempts to save your document in a file with the same name but the extension .dmp. The program displays a message that a serious error has occurred and that all open documents will be closed as *SW* exits.

▶ **To open a .bak file or a .dmp file**

1. From the File menu, choose Open (ALT+F, O).

2. Type the name of the backup file in the File Name box.

3. Choose OK.

---

**Note**   Before you open a .bak or .dmp file, make a copy of the file. Before you work with a .dmp file, use the Windows File Manager to change the extension of the file to .tex.

---

## Closing a Document

Close your document when you've finished the current working session.

▶ **To close a document**

• From the File menu, choose Close (ALT+F, C).
  –or–

• From the control box in the document window, choose Close (CTRL+F4).

–or–

- Double-click the control box in the document window.
  –or–

- From the Window menu, choose Close All (ALT+W, L).
  If you haven't saved your most recent changes, the program displays a warning and asks if you want to save your document.

  - If you do, choose Yes to save the changes before you close the document.
  - If you don't, choose No to discard the changes.

  If you're working on a new document and haven't yet named it, the program displays a warning and asks if you want to save your document.

  - If you do, choose Yes to open the Save As dialog box where you can name and save the document before you close it.
  - If you don't, choose No to discard the document.

## Leaving SW

When you leave *SW*, the program closes any open documents.

▶ **To leave SW**

- From the File menu, choose Exit (ALT+F, X).
  –or–

- From the control box in the *SW* main window, choose Close (ALT+F4).
  –or–

- Double-click the control box in the *SW* main window.
  If you haven't saved your most recent changes to an open document, the program displays a warning and asks if you want to save the document.

  - If you do, choose Yes to save the changes before you leave *SW*.
  - If you don't, choose No to discard the changes.

  If you're working on a new document and haven't yet named it, the program displays a warning and asks if you want to save the document.

  - If you do, choose Yes to open the Save As dialog box where you can name and save the document before you leave *SW*.
  - If you don't, choose No to discard the document.

# 3 Entering and Editing Text

When you open a document in *SW*, the insertion point appears at the top left of the document window. If you've opened a new document, you can begin entering material right away. If you've opened an existing document, you may want to move the insertion point before you add to the existing material.

Entering material in *SW* is just like typing on a typewriter, even when special characters or alphabets are involved. You can move quickly through your *SW* document with the navigation tools. Otherwise, moving around with the mouse or the keyboard is similar to moving around in other computer applications. This chapter explains how to enter text, and Chapter 4 "Entering and Editing Mathematics" explains how to enter mathematics.

## Typing Text

In nearly all styles, *SW* defaults to text. Therefore, when you start the program, the Math/Text button in the main window appears as $\boxed{T}$ . The program assumes that whatever you type is text until you actively change to mathematics. See Chapter 4 "Entering and Editing Mathematics" for an explanation of how to enter mathematics.

▶ **To enter text**

- Type the text from the keyboard.

▶ **To add to text you've already typed**

1. Use the mouse to position the I-beam pointer where you want to begin typing, then click.

   –or–

   Use the arrow keys to move the insertion point where you want to begin typing.

2. Type the new text from the keyboard.

---

**Note**   You may find that you need to type some text items, such as your name or address, again and again. You can store these frequently typed items as fragments and insert them with a minimum of commands. For more information, see Chapter 6 "Using Special Features."

---

## Entering Special Text Characters

You may need to include special text characters and symbols in your documents, especially if you work in languages other than English. In *SW* you can enter special characters and symbols using the mouse or the keyboard. See Chapter 4 "Entering and Editing Mathematics" for information about entering mathematical symbols and characters.

### Typing Accented Characters

Characters accented with special symbols, such as ~ or ^, are common in languages other than English. You can use the mouse to enter an accented character from the Latin-1 and Latin Extended-A Unicode panels on the Symbol toolbar, or you can type a character and add an accent to it.

▶ **To enter a character from a Unicode panel**

1. On the Symbol toolbar,

   - Click 🔲 to open the Latin-1 panel.
     –or–
   - Click 🔲 to open the Latin Extended-A panel.

2. Click the character you want.

▶ **To add an accent to a typed character**

1. Select the character you want to accent or position the insertion point to its right.

2. Add the accent:

   a. On the Standard toolbar, click the Properties button 🔍 .
      –or–
      From the Edit menu, choose Properties (ALT+E, O).
      –or–
      Press CTRL+F5.
   b. Select the accent you want from the Character Properties dialog box.
   c. Choose OK.
      –or–
      Press the keyboard shortcut for the accent you want:

| To enter | Press |
|---|---|
| ^ | CTRL+^ (CTRL+SHIFT+6) |
| ~ | CTRL+~ (CTRL+SHIFT+`) |
| ´ | CTRL+' |
| ` | CTRL+` |
| . | CTRL+. |
| ¨ | CTRL+" (CTRL+SHIFT+') |
| ¯ | CTRL+= |
| → | CTRL+- |

## Typing Punctuation

Besides the standard symbols that appear on your keyboard, you can enter several common punctuation marks and symbols from the Symbol toolbar or from the keyboard.

In addition to the hyphen (-) used to create compound words such as *double-edged* or *brother-in-law*, *SW* recognizes the *en dash* (–) used for number or time ranges, as in 3–8 or April–August, and the *em dash* (—) used to mark parenthetical information or a summary phrase or clause, as shown in the next sentence. You enter all three punctuation marks—the hyphen, the en dash, and the em dash—from the keyboard by pressing the hyphen key (-) one or more times. Similarly, you can use the keyboard to enter curled opening and closing quotation marks, inverted question marks (¿), and inverted exclamation points (¡). Even though you press two or more keystrokes to enter these marks, *SW* interprets them as single symbols.

| To enter | | With the keyboard | With the mouse |
|---|---|---|---|
| - | (hyphen) | Press - | |
| – | (en dash) | Press - two times | Click £ then – |
| — | (em dash) | Press - three times | Click £ then — |
| ¿ | (inverted question mark) | Press ?,' | Click £ then ¿ |
| ¡ | (inverted exclamation point) | Press !,' | Click £ then ¡ |
| " | (curled opening quotation marks) | Press ' twice | Click "" then " |

## Typing ANSI Characters

If your keyboard has a numeric keypad, you can use it to enter ANSI characters. The following table gives the numbers associated with ANSI characters.

▶ **To enter an ANSI character**

1. Hold down the ALT key.

2. On the keypad, type **0** and the number for the character you want.

   For example, to enter ñ, type **0241** on the keypad.

3. Release the ALT key.

**ANSI Characters**

| | | | | | | | | | | | | |
|---|---|---|---|---|---|---|---|---|---|---|---|---|
| 160 | space | 176 | ° | 192 | À | 208 | Ð | 224 | à | 240 | ð |
| 161 | ¡ | 177 | ± | 193 | Á | 209 | Ñ | 225 | á | 241 | ñ |
| 162 | ¢ | 178 | ² | 194 | Â | 210 | Ò | 226 | â | 242 | ò |
| 163 | £ | 179 | ³ | 195 | Ã | 211 | Ó | 227 | ã | 243 | ó |
| 164 | ¤ | 180 | ´ | 196 | Ä | 212 | Ô | 228 | ä | 244 | ô |
| 165 | ¥ | 181 | µ | 197 | Å | 213 | Õ | 229 | å | 245 | õ |
| 166 | ¦ | 182 | ¶ | 198 | Æ | 214 | Ö | 230 | æ | 246 | ö |
| 167 | § | 183 | · | 199 | Ç | 215 | × | 231 | ç | 247 | ÷ |
| 168 | ¨ | 184 | ¸ | 200 | È | 216 | Ø | 232 | è | 248 | ø |
| 169 | © | 185 | ¹ | 201 | É | 217 | Ù | 233 | é | 249 | ù |
| 170 | ª | 186 | º | 202 | Ê | 218 | Ú | 234 | ê | 250 | ú |
| 171 | « | 187 | » | 203 | Ë | 219 | Û | 235 | ë | 251 | û |
| 172 | ¬ | 188 | ¼ | 204 | Ì | 220 | Ü | 236 | ì | 252 | ü |
| 173 | – | 189 | ½ | 205 | Í | 221 | Ý | 237 | í | 253 | ý |
| 174 | ® | 190 | ¾ | 206 | Î | 222 | þ | 238 | î | 254 | Þ |
| 175 | ¯ | 191 | ¿ | 207 | Ï | 223 | ß | 239 | ï | 255 | ÿ |

## Typing Numbers Using the Numeric Keypad

When typing numbers, you may prefer to use the numeric keypad, which has keys and numbers arranged like those on a conventional adding machine. The function of the keypad depends upon the NUM LOCK key:

- With NUM LOCK on, you can type numbers and mathematical symbols using the keypad.

- With NUM LOCK off, you can move the insertion point using the keypad.

  The Status bar displays the current state of the NUM LOCK key.

## Typing Spaces and Breaks

In *SW,* text spacing, line breaks, and page breaks are automatic. As you enter text, the program automatically adds the appropriate amount of space between text items, according to recognized standards of typesetting. The program also creates line and page breaks wherever necessary to fit the material onto the printed page, according to the style you choose for your document.

On the screen, *SW* breaks each line of text whenever necessary to fit the material inside the document window. If you change the size of the window, the program automatically adjusts the on-screen line breaks. The size of the printed page and the size of the document window are not the same, so the line breaks you see on the screen are not necessarily those that will appear in print. To examine the typeset spacing, preview your document. See Chapter 5 "Previewing and Printing Documents" for instructions.

**Tip**    Paragraph marks (¶), which do not print but can be displayed on the screen, show you where paragraphs end. You may find that your work is easier if you have paragraph marks and other nonprinting characters, such as spaces, displayed on the screen. To display nonprinting characters, click  ¶  on the Standard toolbar or choose Invisibles from the View menu (ALT+V, I).

Although spacing is automatic, you may want to add additional horizontal space, vertical space, or line and page breaks at specific points in your document.

## Adding Horizontal Space

As you enter text, the program automatically adds the appropriate amount of horizontal space between text items. You don't have to worry about adding the correct number of spaces after certain punctuation marks, such as periods or colons. In some cases, however, you may want to add additional space between words. Use the Space button on the Math toolbar or the Spacing command on the Insert menu to insert horizontal space.

| Space | Size |
|---|---|
| Em space | width of M |
| 2-em space | width of MM |
| Normal | $\frac{1}{6}$ em |
| Required | $\frac{1}{6}$ em |
| Nonbreaking | $\frac{1}{6}$ em |
| Thin space | $\frac{2}{9}$ em |
| Thick space | $\frac{5}{18}$ em |
| Zero space | 0 em |
| Negative thin space | $-\frac{2}{9}$ em |
| Italic correction | Depends on character to left |

The italic correction, which is inserted at the end of an italicized word that is followed by an upright word, is intended to add extra space so the italicized word does not appear to lean into the upright word, as in this example:

Without italic correction:    *Italicized word* Upright word
With italic correction:    *Italicized word*  Upright word

The correction is automatic, and its size depends on the character to the left of the upright word. For example, after a short letter such as $a$ or $o$ the italic correction is zero. After a letter such as $f$ or $d$ the italic correction is nonzero.

You can also specify a custom horizontal space that is fixed in length or that stretches. If you specify a fixed amount of space, LaTeX will insert that amount when your document is typeset. This line has $\frac{1}{2}$ inch of fixed horizontal space added here        before the rest of the sentence. If you specify stretchy space, LaTeX will determine the exact amount of space that is inserted, based on the stretch factor you specify. A stretch factor of 1 fills the current line. You can fill stretchy space with dots or a line if you set the stretch factor to 1. We've added space with a stretch factor of 1 to the end of this sentence and specified that the space be filled with dots . . . . . . . . . . . . . . . . . . . . . . . . . . . . . . . . . . . . .

In text, if you press SPACEBAR twice in succession or if you press TAB, the program asks you to confirm that you want additional space at the insertion point. You can tell *SW* to add additional horizontal space or to ignore the second space or TAB keypress. Also, you can set the default for how the program handles pressing SPACEBAR twice in succession or pressing TAB. For more information, read about User Setup in Chapter 13 "Customizing SW." When you are entering mathematics, you may need to add additional space in certain circumstances. See Chapter 4 "Entering and Editing Mathematics" and Chapter 6 "Using Special Features" for more information.

---

**Note**    If you want to align text elements, use one of the body paragraph tags (see "Using Body Paragraph Tags" later in this chapter) or insert a table (see Chapter 6 "Using Special Features") instead of adding additional horizontal space or pressing TAB.

---

### ▶ To enter additional horizontal space

1. Position the insertion point where you want additional space.

2. From the Insert menu, choose Spacing, and then choose Horizontal Space (ALT+I, S, H).

   –or–

   On the Math toolbar, click the Space button  .

3. From the Horizontal Space dialog box, choose the space you want.

4. If you want to specify the exact amount of space to be entered

   a. Choose Custom (C).
   b. Specify whether you want the space to be fixed or stretchy.
      • If you specify fixed space, specify the amount of space and the unit of measure.
      • If you specify stretchy space, specify the stretch factor and the fill.
   c. Specify whether you want the space to be inserted always or whether it can be discarded if it falls at the end of a typeset line.

5. Choose OK.

   The program inserts the space you specify. When the Invisibles option is turned on, additional space appears on the screen as a horizontal line. The line doesn't appear in print.

### Adding Vertical Space

Even though *SW* inserts appropriate space between lines, you may want to increase that space in some cases, such as before and after tables or graphics.

▶ **To enter additional vertical space**

1. Position the insertion point where you want additional space.

2. From the Insert menu, choose Spacing, and then choose Vertical Space (ALT+I, S, V).

3. From the Vertical Space dialog box, choose the space you want:

| Space | Size |
|---|---|
| Small skip | about $\frac{5}{16}$ inch between lines in a 12-point style |
| Medium skip | about $\frac{3}{8}$ inch between lines in a 12-point style |
| Big skip | about $\frac{7}{16}$ inch between lines in a 12-point style |
| Strut | about $\frac{3}{16}$ inch between lines in a 12-point style |
| Math strut | about $\frac{7}{32}$ inch between lines in a 12-point style |

4. If you want to specify the exact amount of space to be entered

   a. Choose Custom Space (C).
   b. Specify the amount of space and the unit of measure.
   c. Specify whether you want the space to be inserted always or whether it can be discarded if it falls at the end of a typeset page.

5. Choose OK.

   When the Invisibles option is turned on, the space appears in the document window as a green vertical line. The line doesn't print.

## Adding Line and Page Breaks

Line and page breaks are automatic, so you don't need to press ENTER at the end of every line. Instead, press ENTER only when you want to start a new paragraph. You can add a line break to start a new line in the middle of a paragraph, and you can override automatic page breaks to start a new page at a specific point in your document.

If you press ENTER twice in succession, the program displays a message that you are trying to enter an empty paragraph. You can tell *SW* to ignore the empty paragraph or to add additional vertical space. You can also set the default for how the program responds when you press ENTER twice. For more information, read about User Setup in Chapter 13 "Customizing SW."

▶ **To start a new paragraph**

• Press ENTER.
  The program ends the current paragraph and moves the insertion point to the leftmost position in the first line of the new paragraph.

▶ **To start a new line in the same paragraph**

1. Position the insertion point where you want the new line to begin.

2. From the Insert menu, choose Spacing, and then choose Break (ALT+I, S, B).

3. From the Break dialog box, choose the line break you want:

| Break | Use |
|---|---|
| Newline | Start a new line at the break |
| Linebreak | Start a new line at the break and fully justify the text on the line |
| Custom newline | Start a new line at the break after a specified amount of vertical space |

4. If you want to specify the amount of space to skip before beginning a new line

    a. Choose Custom newline (C).
    b. Specify the amount of space and the unit of measure.
    c. Specify whether you want the space to be inserted always or whether it can be discarded if it falls at the end of a typeset page.

5. Choose OK.

▶ **To start a new page**

1. Position the insertion point where you want the new page to begin.

2. From the Insert menu, choose Spacing, and then choose Break (ALT+I, S, B).

3. From the Break dialog box, choose the page break you want:

| Break | Use |
|---|---|
| Newpage | Start a new page and a new paragraph at the break |
| Pagebreak | Start a new page after the line on which the break occurs |

4. Choose OK.

The program inserts the page break you specify. When the Invisibles option is turned on, a page break appears on the screen as a horizontal line; the line doesn't appear in print.

## Using Text and Body Paragraph Tags

Although the typesetting style determines how your document will look in print, you can apply text and body paragraph tags to add emphasis to text selections within a paragraph and to set off pieces of text from the main part of a paragraph. You can apply tags from the Tag toolbar, with the Apply command, and from the keyboard with the function key assignments. Chapter 10 "Applying Styles to Documents" contains more information about typesetting styles

▶ **To apply a tag from the tag popup lists**

1. Click the popup box containing the tag you want.

2. Click the tag.

▶ **To apply a tag using the Apply command**

1. From the Tag menu, choose Apply (ALT+T, A).

2. Select the tag you want.

3. Choose OK.

▶ **To apply a tag using the function keys**

• Press the function key assigned to the tag you want.
  When you install *SW*, certain tags are assigned to the function keys on the keyboard:

| Key | Tag |
|-----|-----|
| F2 | Remove Item Tag |
| F3 | Body Text |
| F4 | Normal |
| F5 | Bold |
| F6 | Emphasize |
| F7 | Numbered item |
| F8 | Bullet item |
| F9 | Calligraphic |
| F11 | Section head |
| F12 | Subsection head |

You can change the key assignments with the Function Keys command on the Tag menu. See Chapter 13 "Customizing SW."

## Using Text Tags

You can add emphasis to text within a paragraph by applying text tags such as bold, italic, or small caps. Some styles use content-oriented tag names such as Define and Emphasize instead of appearance-oriented names such as Bold and Italic. Most non-Style Editor styles have text tags for the standard LaTeX text sizes, such as tiny, large, and huge. Text tags affect the way tagged text in your document appears in print and, depending on the style, on the screen. You can apply text tags to the text you type at the insertion point or to a selection of text you have already entered.

---

**Note**    Although many of the text tags have names that indicate an appearance, you should think of tags as providing shades of meaning to words by emphasizing them. In mathematics, the Bold tag is often used to indicate vectors and matrices, while the Calligraphic tag is used for sets. Because these tags apply meaning rather than appearance, it makes little sense to accumulate them. *SW* always uses only the most recently applied tag.

---

▶ **To apply a text tag to a selection**

1. Select the text you want to tag.

2. Choose the tag.

▶ **To apply a text tag at the insertion point**

1. Choose the tag.

2. Type the text to be tagged.

3. Choose the (Normal) text tag.

▶ **To remove a text tag**

1. Select the tagged text.

2. Choose the (Normal) text tag.

## Using Body Paragraph Tags

In nearly all styles, all text in an *SW* document is automatically tagged as Body Text. You can use body paragraph tags to center a paragraph or a quotation set off from the main body of the document. You can also use the body verbatim tag so that text appears in print unformatted and exactly as it is typed. The body verbatim tag is useful for depicting text such as sections of code, with line breaks, spacing, and special characters preserved.

You can also apply a body paragraph tag to make a paragraph a heading; the style you choose determines which section heading tags are available. See Chapter 11 "Structuring Documents" for details.

▶ **To apply a body paragraph tag**

1. Place the insertion point in the paragraph you want to set off from the body of the document.

2. Choose the tag you want.

---

**Note**    If you want to apply a body paragraph tag to several consecutive paragraphs, make a selection that starts anywhere in the first paragraph and ends anywhere in the last paragraph. Then apply the tag. *SW* applies the tag to any paragraph that contains a part of the selection, even if the entire paragraph was not selected.

---

▶ **To remove a body paragraph tag**

1. Place the insertion point in the paragraph.

2. Choose the Body Text tag.

▶ **To center a paragraph**

1. Type the paragraph.

2. Choose the tag.

   - Choose the Body Center tag to center a short, one-line body text paragraph.
   - Choose the Body Quote tag to set off a short, one-paragraph quotation.
   - Choose the Body Quotation tag to set off a multiparagraph quotation.

---

**Tip**    When you want to center a heading, avoid using the Body Center tag. Instead, type the heading and apply a section tag (see Chapter 10 "Applying Styles to Documents"), then choose a typesetting style that centers that level of heading.

---

# Editing Text

In *SW* you edit your document much as you would in any other word processing application: by adding or changing information, by changing the properties of existing text, or by deleting, copying, or moving information. You can use the BACKSPACE or DELETE key to correct typing mistakes as you work and to remove information permanently from your document. You can also cut, copy, and paste information by applying standard clipboard operations with the keyboard or the mouse or by using a mouse technique called *drag and drop*.

## Editing Text Properties

Some text changes, such as adding accents, involve editing the properties of characters.

▶ **To edit the properties of a character**

1. Select the character or place the insertion point to its right.

2. Choose Properties using one of these commands:

- On the Standard toolbar, click the Properties button [image] .
  –or–
- From the Edit menu, choose Properties (ALT+E, O).
  –or–
- Press CTRL+F5.

The program opens a context-sensitive Properties dialog box so you can change the character.

## Deleting and Reinserting Information

You can delete information permanently from your document. If you change your mind, you can undo the most recent deletion, but only if you've made no other editing changes since that deletion and if the Undo Deletion command on the Edit menu is active. Otherwise, you can't reinsert the deleted information unless you retype it.

▶ **To delete a selection permanently**

- Press DELETE or BACKSPACE.
  –or–
- From the Edit menu, choose Delete (ALT+E, D).

▶ **To undo the last deletion**

- On the Standard toolbar, click the Undo button [image] .
  –or–
- From the Edit menu, choose Undo Deletion (ALT+E, U).
  –or–
- Press CTRL+Z.

## Cutting, Copying, and Pasting Information

Using the Edit menu, the buttons on the Standard toolbar, or the corresponding keyboard equivalents, you can cut or copy a selection to the clipboard and move, or *paste*, it to a new location. Your selection remains in the clipboard until it is overwritten by the next clipboard operation.

If you place text on the clipboard from within *SW* and then paste it into an *SW* document, the information is pasted into your document in *SW* internal format. If you place text on the clipboard from another application, the information is pasted into your document as unformatted text. If the clipboard contains a graphic with a `.wmf` or `.bmp` file extension, the graphic is pasted into your document in that same format.

### ▶ To cut a selection to the clipboard

- On the Standard toolbar, click the Cut button ✂.
  –or–
- From the Edit menu, choose Cut (ALT+E, T).
  –or–
- Press CTRL+X.
  *SW* deletes the selection from your document and moves it to the clipboard.

### ▶ To copy a selection to the clipboard

- On the Standard toolbar, click the Copy button ⧉.
  –or–
- From the Edit menu, choose Copy (ALT+E, C).
  –or–
- Press CTRL+C.
  *SW* makes a copy of the selection on the clipboard, leaving the selection in place in your document.

### ▶ To paste the contents of the clipboard at the insertion point

- On the Standard toolbar, click the Paste button ▣.
  –or–
- From the Edit menu, choose Paste (ALT+E, P).
  –or–
- Press CTRL+V.
  *SW* pastes the contents of the clipboard into your document at the insertion point. The information remains on the clipboard until you cut or copy other information.

## Using Paste Special

With the Paste Special command, you can specify the format in which text on the clipboard is pasted into your document. You can specify that the text is to be pasted as *SW* Internal format or as unformatted text, in which case *SW* does not attempt to interpret it. The Paste Special command is useful for inserting information copied from a line editor so that it appears in your *SW* document with the line breaks as they appear in the line editor. Font specifications made in other applications are not preserved. These formats are available:

| Format | Effect |
|---|---|
| SW Internal | Interpret the clipboard information for *SW* and paste the interpreted information into the document |
| Text, Paste Text as Unformatted Text | Paste the clipboard information exactly as it appears |
| Text, Paste Text as SW Internal | Interpret the clipboard information for *SW* and paste the interpreted information into the document |

For example, if you're working in *SW*, copy the Greek letter $\alpha$ to the clipboard, and then use **Paste Special** to paste it into your document in *SW* Internal format, the program inserts $\alpha$ at the insertion point. If you paste the $\alpha$ into your document as Unformatted Text, at the insertion point the program inserts

$$\$\backslash alpha\$$$

Similarly, you can copy unformatted text from another application and have the program interpret it in *SW* Internal format. If you're working in another application and place the unformatted text **$\alpha$** on the clipboard, then use Paste Special to paste the clipboard contents into your document as Text in *SW* Internal format, the program inserts $\alpha$.

Please note, however, that incorrectly formed expressions pasted to the clipboard can create problems when the program tries to interpret those expressions as SW Internal format. Note also that although *SW* Internal format agrees with TeX and LaTeX for many symbols, it does not agree for many other elements, including matrices and tables.

---

**Tip**    *SW* filters and treats clipboard text in much the same way as the ASCII and ANSI document filters. If you want to preserve line ends and spacing from the clipboard, paste into a paragraph that you've tagged as Body Verbatim.

---

▶ **To specify the format in which information is pasted from the clipboard**

1. From the **Edit** menu, choose **Paste Special** (ALT+E, S).

2. Select the format you want.

3. Choose **OK**.

*SW* pastes the contents of the clipboard into your document at the insertion point, using the format you specified. The selection remains on the clipboard until you cut or copy other information.

## Using Drag-and-Drop Techniques

You can use drag-and-drop techniques to delete, copy, and move selections with the mouse. When you use drag-and-drop techniques to delete information, the information is deleted permanently. You can reinsert it immediately with the Undo Deletion command, but if you make any other editing choices before you try to reinsert the deleted information, you must retype it.

The drag-and-drop techniques are assigned to the mouse buttons by default. You can change the default assignments in the User Setup dialog box. See Chapter 13 "Customizing SW" for more information.

▶ **To delete a selection**

1. Select the information you want to delete.

2. Without pressing any of the mouse buttons, move the mouse pointer within your selection.

3. Press and hold down the left mouse button.

   The mouse pointer changes to the scissors pointer, indicating that your selection will be cut from its current position.

4. While holding down the mouse button, drag the mouse pointer to the side of the *SW* window, outside the text region. The most convenient place is the scroll bar.

5. Release the mouse button.

   The program permanently removes the selection from your document.

▶ **To move a selection**

1. Select the text you want to move.

2. Without pressing any of the mouse buttons, move the mouse pointer within your selection.

3. Press and hold down the left mouse button.

   The mouse pointer changes to the scissors pointer, indicating that your selection will be cut from its current position.

4. While holding down the mouse button, drag the mouse pointer to the place where you want the selection to appear.

   The insertion point follows the mouse pointer as you do this, showing you accurately where your selection will appear.

5. Release the mouse button.

   The program moves your selection from its former position to the place where you released the mouse button.

▶ **To copy a selection**

1. Select the text you want to copy.

2. Without pressing any of the mouse buttons, move the mouse pointer within your selection.

3. Hold down the CTRL key, and then press and hold down the left mouse button.

   The mouse pointer changes to the copy pointer, indicating that the selection will be copied.

4. While holding down the CTRL key and the mouse button, drag the mouse pointer to the place where you want the selection to be copied.

   The insertion point follows the mouse pointer as you do this, providing an accurate guide to where your selection will appear.

5. Release the CTRL key and the mouse button.

   The program copies your selection to the place where you released the mouse button.

▶ **To cancel a drag and drop operation while it is in progress**

1. Keep holding down the mouse button and return the insertion point to within the selection.

2. Release the mouse button.

# 4 Entering and Editing Mathematics

Entering mathematics in $SW$ is straightforward. You can focus on content when you're entering mathematics, just as when you're entering text. You use familiar mathematical notation to enter mathematical characters, symbols, and objects into your document, and you use simple commands to create displayed or in-line mathematics. You edit your mathematics with the usual operations of select, cut, copy, move, and delete.

This chapter explains how $SW$ treats mathematics and gives instructions for entering and editing mathematical information. The online Tutorial exercises contain step-by-step instructions for entering specific mathematical expressions.

## Text and Mathematics

In nearly all styles, $SW$ defaults to text. When you start $SW$, the Math/Text button in the main window appears as $\boxed{\text{T}}$. This means that the insertion point is in text. The program assumes that whatever you enter is text until you tell it you want to enter mathematics. Then the button appears as $\boxed{\text{M}}$.

## Changing from Text to Mathematics or Mathematics to Text

The Math/Text button on the Math toolbar indicates whether the insertion point is in text or mathematics. You can move from text to mathematics and back again three different ways. Use the method you prefer.

▶ **To start mathematics**

- Click the Math/Text button to switch to $\boxed{\text{M}}$.
  –or–
- Press CTRL+M.
  –or–
- Press CTRL+T.
  –or–
- From the Insert menu, choose Math (ALT+I, M).

▶ **To return to text**

- Click the Math/Text button to switch to [ T ].

  –or–

- Press CTRL+T.

  –or–

- Press CTRL+M.

  –or–

- From the Insert menu, choose Text (ALT+I, T).

  You can also customize *SW* to toggle between text and mathematics in several other ways. See Chapter 13 "Customizing SW" for more information.

## Understanding the Differences Between Mathematics and Text

*SW* treats text and mathematics differently. On the screen, *SW* uses color to distinguish between text and mathematics. The default color for text is black; the default color for mathematics is red. Math names for functions, operators, and variables are shown in gray. The program also uses these conventions:

When *SW* is in text

- Characters are normally upright.
- The SPACEBAR is used to end words. Unless you customize the use of the SPACEBAR (see Chapter 13 "Customizing SW"), pressing SPACEBAR a second time has no effect. The program automatically provides the correct text spacing when the document is printed.
- The ENTER key is used to end paragraphs. Pressing ENTER a second time has no effect unless you customize its use as described in Chapter 13 "Customizing SW."

When *SW* is in mathematics

- Alphabetic characters are italicized.
- Numbers and math names are upright.
- The right quotation mark key enters a superscripted prime character.
- Space is automatically inserted around operators such as + and relations such as =.
- The SPACEBAR advances the insertion point to the next object. *SW* automatically provides the correct spacing in formulas.

  *SW* automatically formats mathematics and text correctly when you print your document.

## Creating In-Line and Displayed Mathematics

You can create *in-line* or *displayed* mathematics. In-line mathematics appears in a line of text. Any operators are sized to fit within the line, and limits are positioned to the right of operators and functions. The numerators and denominators of fractions are set in small type; for example,

$$\ldots \text{ and let } S = \sum_{n=1}^{k} \frac{a_n + b_n}{c_n}. \text{ Then } S \text{ is } \ldots$$

Displayed mathematics appears centered on a separate line and set off from the text above and below by additional space. Operators within the display are larger, limits are positioned above and below operators and functions, and the numerators and denominators of fractions are set in full-size type; for example,

$\ldots$ and let

$$S = \sum_{n=1}^{k} \frac{a_n + b_n}{c_n}$$

Then $S$ is $\ldots$

By default, $SW$ doesn't number equations when you preview or print your document, but you can change the default from the **User Setup** dialog box. See Chapter 13 "Customizing SW" for more information. You can edit the properties of a mathematics display to add an optional label. When the document is printed, the label becomes an automatically generated equation number. If you supply a unique name, or *key,* for the display, you can use the key to create a *cross-reference* to the equation from elsewhere in the text. When the document is printed, the reference is replaced by the equation number.

---

**Note**    To label multiline displays, see "Entering Multiline Displays" later in this chapter.

---

### ▶ To create a mathematics display

1. On the Math toolbar, click the Display button  .

    −or−

    From the **Insert** menu, choose **Display** (ALT+I, D).

    −or−

    Press CTRL+D.

2. Enter your mathematics.

    If you have your view set to display matrix lines, the display is outlined with a box. This box does not print.

### ▶ To display in-line mathematics

1. Select the mathematics to be displayed.

2. On the Math toolbar, click the Display button  .

–or–

From the Insert menu, choose Display (ALT+I, D).

–or–

Press CTRL+D.

▶ **To change displayed mathematics to in-line mathematics**

- Place the insertion point to the left of and outside the display and press DELETE.
  –or–
- Place the insertion point to the right of and outside the display and press BACKSPACE.

▶ **To add a label to a display**

1. Edit the properties of the display:

   - Double-click outside and to the left or right of the display.
     –or–
   - With the insertion point to the right of and outside the display, choose Properties:

     - On the Standard toolbar, click the Properties button [icon].
       –or–
     - From the Edit menu, choose Properties (ALT+E, O).
       –or–
     - Press CTRL+F5.

2. Type a key for the label if you want to refer to it from elsewhere in the text.

   The program substitutes the equation number for the reference when the document is printed.

3. Set the numbering for the label:

   - Choose None for an unnumbered display.
   - Choose Auto for an automatically generated display number.
   - Choose Custom to enter your own label instead of the automatically generated number.

4. Check Suppress Annotation to remove enclosures from around the equation number.

   The presence of enclosures is determined by the style you choose for the document. You can suppress the enclosures in AMS styles only.

5. Choose OK.

The program displays a gray box with a # to the right of the display to indicate that a number will be generated when you print or preview your document. The form of this number is determined by the typesetting style you choose for the document. If you specified a key, the name appears in the gray button like this:  # (key) . If you supplied a custom label, it is displayed in the button in place of the number:  label (key) . For information on how to refer to the equation in the text, see Chapter 6 "Using Special Features."

## Entering and Editing Symbols and Characters

In *SW*, you can enter and edit mathematics symbols and characters as naturally as you enter and edit text.

### Entering Symbols and Characters

You can enter symbols and characters directly from the Symbol toolbar, the Common Symbols toolbar, and the keyboard. Or you can enter text and then convert it to mathematics.

▶ **To enter symbols and characters directly**

- Use the keyboard to enter symbols and characters with the keyboard shortcuts and key prefixes.
  See Appendix B "Keyboard Shortcuts and Key Prefixes" for a list of the shortcuts and prefixes.
  −or−

- Use the mouse to enter symbols and characters from the Common Symbols toolbar or the symbol panels shown on page 56.

▶ **To convert text to mathematics**

1. Select the text you want to convert to mathematics.

2. Choose mathematics.

   The Math/Text button appears as  M , the selection is converted to mathematics, and all ordinary spaces in the selection are removed.

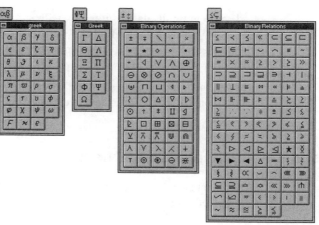

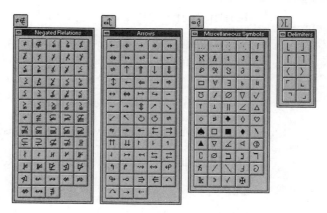

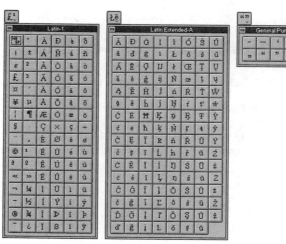

## Editing Symbols and Characters

You can edit symbols and characters by adding mathematical accents to them and by negating them. For both operations, use the **Properties** command or the keyboard shortcuts. If you use the **Properties** command, *SW* displays the **Character Properties** dialog box so you can accent or negate a character or symbol. Your symbols and characters can carry these mathematical accents:

$$\hat{a} \; \breve{a} \; \tilde{a} \; \acute{a} \; \grave{a} \; \dot{a} \; \ddot{a} \; \check{a} \; \bar{a} \; \vec{a}$$

▶ **To edit the properties of a symbol or character**

- On the Standard toolbar, click the Properties button ⊡.
  –or–
- From the Edit menu, choose **Properties** (ALT+E, O).
  –or–
- Press F5.

▶ **To add a mathematical accent**

1. Enter the symbol or character to be accented.

2. Use the **Properties** command to open the **Character Properties** dialog box.

   –or–

   Choose the keyboard shortcut for the accent you want:

   | For this accent | Press |
   |:---:|:---|
   | ^ | CTRL+^ |
   | ~ | CTRL+~ |
   | ´ | CTRL+' |
   | ` | CTRL+' |
   | . | CTRL+. |
   | .. | CTRL+" |
   | − | CTRL+= |
   | → | CTRL+- |

   There are no keyboard shortcuts for $\breve{a}$ and $\check{a}$.

▶ **To remove a mathematical accent**

1. Position the insertion point to the right of the accented character or symbol.

2. Use the **Properties** command to open the **Character Properties** dialog box.

3. Click the selected accent to deselect it.

4. Choose **OK**.

► **To negate a character or symbol**

1. Enter the symbol or character, or position the insertion point to the right of the symbol or character to be negated.

2. Press CTRL+N.

   –or–

   Use the Properties command to open the Character Properties dialog box, check Negate, and then choose OK.

► **To remove negation from a symbol or character**

1. Position the insertion point to the right of the negated character.

2. Use the Properties command to open the Character Properties dialog box.

3. Uncheck Negate.

4. Choose OK.

## Entering and Editing Mathematical Objects

You can enter mathematical objects from the Math toolbar and from the Insert menu. You can also select the most common objects using the keyboard shortcuts. Some mathematical objects, such as fractions and radicals, have defaults that you can customize. See Chapter 13 "Customizing SW" for more information.

## Entering Objects with Templates

When you enter a mathematical object, *SW* inserts a template representing the object and places the insertion point within the template so you can complete the object. For example, when you select the Fraction command or click the Fraction button, *SW* inserts a fraction template that shows a bar between two small boxes. The boxes represent the numerator and the denominator. The insertion point appears in the numerator box, like this:

$$\frac{\Phi}{\Box}$$

Depending on the object, a template may have several small input boxes.

► **To move from box to box within a template**

- Use the TAB key or the ARROW keys.
  –or–
- Point to the box you want and click.

*SW* expands templates as necessary. Brackets and radicals expand both horizontally and vertically to encompass their contents; fraction bars extend as far as necessary to encompass the longest string of characters within the fraction; and matrix cells expand as needed.

Until you move the insertion point outside the template, *SW* will include anything you enter as part of the object. When you have completed the object, click to the right of the template with the mouse or press RIGHT ARROW or SPACEBAR repeatedly until the insertion point moves outside the template. The individual mathematical objects are discussed later in this chapter.

## Applying and Removing Templates

You can apply a template to a selection and, conversely, a selection to a template.

▶ **To apply a template**

1. Select the information to which you want to apply the template.

2. Select the template.

The program places your selection within the template. If you apply a template with several input boxes, the program places the selection in the first one.

▶ **To make a selection part of a template**

1. Insert the template.

2. Select the mathematics you want to insert in the template.

3. Use drag and drop to copy or move the selection to the template.

   –or–

1. Make the selection.

2. Copy or cut the selection to the clipboard.

3. Insert the new template.

4. Paste the selection into the template.

▶ **To remove a template without deleting its contents**

1. Place the insertion point to the left of the template.

2. Press DELETE.

---

**Note**    If more than one input box in the template is filled, *SW* deletes both the template and its contents. If only the first input box is filled, *SW* deletes the template (except radicals and labels) without deleting its contents.

---

## Editing Mathematical Objects

Edit your mathematical objects by typing; by editing the properties of the object; or by using the same select, cut, copy, paste, and delete operations you use to edit text (see Chapter 3 "Entering and Editing Text"). Some objects have special editing features; they are discussed later in this chapter.

▶ **To edit the properties of an object**

1. Select the object or place the insertion point to its right.

2. Choose **Properties**:

   - On the Standard toolbar, click the Properties button .
     –or–
   - From the Edit menu, choose **Properties** (ALT+E, O).
     –or–
   - Press CTRL+F5.

3. Edit the object.

–or–

1. Double-click the object.

   Be careful to double-click the object itself and not one of its fields. For example, you should double-click the bar of a fraction, not the numerator or denominator.

2. Edit the object.

## Entering Fractions and Binomials

*SW* treats fractions and binomials similarly.

### Fractions

▶ **To enter a fraction**

- On the Math toolbar, click the Fraction button .
  –or–
- From the Insert menu, choose **Fraction** (ALT+I, F).
  –or–
- Type CTRL+F.
  Additional keyboard shortcuts for entering fractions appear in Appendix B "Keyboard Shortcuts and Key Prefixes."

### In-Line and Displayed Fractions

The numerator and denominator of an in-line fraction are normally set in a small-size font; for example,

$$\frac{a+b}{2}$$

When the fraction is displayed, the numerator and denominator are normally set in a full-size font; for example,

$$\frac{a+b}{2}$$

You can override the defaults to achieve effects like these:

$$\frac{x}{1+\frac{x}{2}} \quad \frac{x}{1+\dfrac{x}{2}}$$

▶ **To override the size defaults**

1. Select the fraction or place the insertion point to its right.

2. Choose **Properties**.

3. Set the Size option in the **Fraction Properties** dialog box to **Big** or **Small**.

▶ **To return to the size defaults**

1. Select the fraction or place the insertion point to its right.

2. Choose **Properties**.

3. In the Size options area of the **Fraction Properties** dialog box, choose **Auto**.

### Binomials

Binomials, the Legendre symbol, and Euler's numbers are treated as *generalized fractions*:

Binomial $\qquad \binom{n}{k}$

Legendre symbol $\qquad \left(\frac{a}{b}\right)$

Euler's number $\qquad \left\langle\frac{a}{b}\right\rangle$

▶ **To enter a binomial or generalized fraction**

- On the Math toolbar, click the Binomial button ⊞ .
  −or−
- From the Insert menu, choose **Binomial** (ALT+I, B).

## Entering Radicals

▶ **To enter a radical**

- On the Math toolbar, click the Radical button $\boxed{\sqrt{\square}}$ .
  –or–
- From the Insert menu, choose Radical (ALT+I, R).
  –or–
- Press CTRL+R.
  Additional keyboard shortcuts for entering a radical appear in Appendix B "Keyboard Shortcuts and Key Prefixes."

▶ **To add a root to a radical**

1. Select the radical or place the insertion point to its right.

2. Choose Properties.

   The program displays the Radical Properties dialog box.

3. Choose the template that has a root.

4. Choose OK.

5. Enter the root.

   –or–

1. Place the insertion point under the radical.

2. Press TAB.

3. Enter the root.

▶ **To enclose an existing expression in a radical**

1. Select the expression.

2. Enter a radical.

▶ **To remove a radical from an expression**

1. Place the insertion point to the left of the radical.

2. Press DELETE.

   –or–

1. Place the insertion point to the right of the radical.

2. Press BACKSPACE.

## Entering Subscripts and Superscripts

In *SW*, you can enter subscripts and superscripts using the menu commands or the icons. You can also create subscripts and superscripts by applying the templates to selections. See "Entering Operators" later in this chapter for details about creating multiline subscripts and superscripts.

### ▶ To enter a subscript

- On the Math toolbar, click the Subscript button $\boxed{N_x}$.
  –or–
- From the Insert menu, choose Subscript (ALT+I, B).
  –or–
- Press CTRL+L.
  Appendix B "Keyboard Shortcuts and Key Prefixes" lists additional keyboard shortcuts for entering a subscript.

### ▶ To enter a superscript

- On the Math toolbar, click the Superscript button $\boxed{N^x}$.
  –or–
- From the Insert menu, choose Superscript (ALT+I, P).
  –or–
- Press CTRL+H.
  Appendix B "Keyboard Shortcuts and Key Prefixes" lists additional keyboard shortcuts for entering a superscript.

  If the insertion point is to the right of a superscript, *SW* interprets a new command to enter a superscript as a command to move the insertion point into the existing superscript. Similarly, if the insertion point is to the right of a subscript, *SW* interprets a new command to enter a subscript as a command to move the insertion point into the existing subscript.

### ▶ To apply a subscript or superscript template to a selection

1. Select the mathematics you want to be a subscript or a superscript.

2. Enter a subscript or superscript.

   The program places your selection in the template you selected.

▶ **To create an expression with a simultaneous subscript and superscript**

1. Enter the expression.

2. Enter the subscript.

3. Press TAB.

4. Enter the superscript.

   –or–

1. Enter the expression.

2. Enter the subscript.

3. Move the insertion point outside the subscript and to the right.

4. Enter the superscript.

---

**Tip**    With either method, you can enter the superscript first and then create the subscript.

---

## Entering Operators and Limits

*SW* provides these operators:

### Entering Operators

▶ **To enter an operator**

- On the Math toolbar, click the Operator button  $\boxed{\Sigma\!\int}$  .
  –or–

- From the **Insert** menu, choose **Operator** (ALT+I, O).

  Both methods open the **Operator** dialog box from which you can select the operator you want. Certain operators have several entry options:

| To enter | Choose | or Press |
|----------|--------|----------|
| $\sum$ | $\boxed{\Sigma}$ | CTRL+S, S |
| $\int$ | $\boxed{\int}$ | CTRL+I or CTRL+S, I |
| $\prod$ | | CTRL+P or CTRL+S, P |
| $\oplus$ | | CTRL+S, SHIFT+S |
| $\otimes$ | | CTRL+S, T |

## Entering Limits

You add limits to operators using the **Subscript** and **Superscript** commands. When an operator is in-line, it is small and the limits are set to the right:

$$\sum_{i=1}^{n} a_i$$

When an operator is displayed, it is large and the limits are set above and below:

$$\sum_{i=1}^{n} a_i$$

The exception is the integral operator: the limits are set to the right in both cases.

You can override these defaults at the time of entry or by editing the properties of the operator later. We recommend that you choose the defaults (Auto for both Limit Position and Size) wherever possible.

You can create multiline limits in operators:

$$\sum_{\substack{0 \le i \le m \\ 0 < j < n}} P(i,j)$$

▶ **To enter a multiline limit**

1. Enter the first line of the limit as a subscript or superscript to the operator.

2. With the insertion point at the end of the limit, press ENTER.

3. Enter the next line of the limit.

Multiline limits can be edited in the same way you edit column vectors. See "Entering and Editing Matrices and Vectors" later in this chapter for details.

## Entering and Editing Brackets

*SW* has two kinds of brackets: standard brackets, which are available from the keyboard, and expanding brackets, which are available from the **Insert** menu and the Math toolbar.

Standard brackets, which can be entered singly, don't change in size. Expanding brackets, which are always entered in pairs, are elastic. They expand or contract to match their contents. *SW* provides these symbols for use as left or right members of expanding bracket pairs:

$$( ) [ ] \{ \} \langle \rangle \lfloor \lceil \rceil \mid \parallel / \backslash \updownarrow \Updownarrow \uparrow \Uparrow \downarrow \Downarrow$$

When you need a single expanding bracket, such as to contain a set of equations, you complete the pair of brackets with an *empty bracket*. Empty brackets do not appear in print, but they are shown on the screen and in the **Brackets** dialog box as vertical dotted lines.

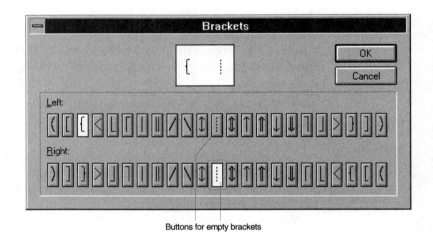

Buttons for empty brackets

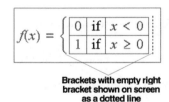

**Brackets with empty right bracket shown on screen as a dotted line**

▶ **To enter expanding brackets**

- On the Math toolbar, click the Brackets button **[()]**.
  –or–
- From the **Insert** menu, choose **Brackets** (ALT+I, K).

Both methods open the **Brackets** dialog box so you can select the pair of expanding brackets you want. Certain bracket combinations have special buttons or keyboard shortcuts:

| To enter | Choose | or Press |
|----------|--------|----------|
| ( ) | (□) | CTRL+9 or CTRL+( or CTRL+0 or CTRL+) |
| [ ] | [□] | CTRL+[ or CTRL+] |
| { } | | CTRL+{ or CTRL+} |
| \| \| | | CTRL+\| |

▶ **To enclose an existing expression in expanding brackets**

1. Select the expression.

2. Enter the bracket.

▶ **To remove expanding brackets from an expression**

- Place the insertion point to the left of either bracket and press DELETE.
  −or−
- Place the insertion point to the right of either bracket and press BACKSPACE.

▶ **To change bracket symbols**

1. Select the bracket, including its contents.

   −or−

   Place the insertion point to the right of the right bracket.

2. Choose Properties to open the Brackets Properties dialog box.

3. Select the symbols you want for the right and left brackets.

4. Choose OK.

## Entering and Editing Matrices and Vectors

A matrix is a two-dimensional array:

$$\begin{matrix} a & b \\ c & d \end{matrix}$$

A vector is a matrix with one column or one row:

$$\begin{matrix} a \\ b \end{matrix} \qquad c \quad d$$

## Entering Matrices

### ▶ To enter a matrix

- On the Math toolbar, click the Matrix button ⊞ .
  –or–
- From the Insert menu, choose Matrix (ALT+I, X).
  Both methods open the Matrix dialog box.

---

**Tip**   *SW* has a keyboard shortcut for one special matrix, the 2×2 matrix: CTRL+A.

---

You can outline the cells of a matrix by choosing Matrix Lines from the View menu. These outlines do not print and are intended only as an editing guide. Most matrices are enclosed in brackets:

$$\begin{pmatrix} \alpha & \beta \\ \gamma & \delta \end{pmatrix}$$

### ▶ To add brackets to a matrix

1. Select the matrix.

2. On the Math toolbar, click the Parentheses button (□) .

   –or–

   On the Math toolbar, click the Square Brackets button [□] .

   –or–

   Press CTRL+( or CTRL+) or CTRL+[ or CTRL+].

## Editing Matrices

Options for editing matrices are somewhat more varied than for other mathematical objects. You can use commands on the Edit menu to add rows and columns and to edit the properties of a matrix. You can delete rows and columns without special commands.

## Deleting and Adding Rows and Columns

You add rows and columns to a matrix from within the matrix.

### ▶ To add rows or columns to a matrix

1. Place the insertion point anywhere in the matrix, but not within another mathematical object.

2. From the Edit menu

   - Choose Insert Row(s) (ALT+E, W).
     –or–
   - Choose Insert Column(s) (ALT+E, I).

Depending on your choice, the program opens the Insert Rows or the Insert Columns dialog box.

3. In the dialog box, select the number of rows or columns you want to add.

4. Enter the position where you want the addition to be made.

   The arrow on the sample matrix moves to the point you indicate.

5. Choose OK.

You can delete the contents of a cell or cells, the contents of a row or column, and the row or column itself without using the Properties command. Note, however, that if you select and delete the contents of an entire row or column in a single operation, *SW* deletes the row or column from the matrix.

To delete the contents of a row or column but keep the row or column in the matrix, delete the contents of at least one cell in a separate operation. *SW* erases the contents of the cells but leaves the input boxes for you to refill.

### ▶ To delete the contents of a cell

1. Select the contents of the cell.

2. Press either DELETE or BACKSPACE.

   The program deletes the contents of the cell and replaces it with an empty input box for you to refill.

### ▶ To delete an entire row or column

1. Select the entire row or column.

2. Press DELETE.

   –or–

   From the Edit menu, choose Delete (ALT+E, D).

### Editing Matrix Properties

You can edit the properties of a matrix to align the contents of the columns; each column can be aligned separately.

You can also align the matrix with the surrounding text. You can set the baseline on the top row, at the vertical center, or on the bottom row of the matrix:

$$\text{Baseline on top row:} \quad \begin{matrix} 1 & 2 \\ 3 & 4 \end{matrix}$$

$$\text{Baseline at center:} \quad \begin{matrix} 1 & 2 \\ 3 & 4 \end{matrix}$$

$$\text{Baseline on bottom row:} \quad \begin{matrix} 1 & 2 \\ 3 & 4 \end{matrix}$$

### ▶ To edit the properties of a matrix

1. Select the matrix or place the insertion point to its right.

2. Choose **Properties**.

3. Make your choices from the **Matrix Properties** dialog box.

4. Choose **OK**.

### Editing Vectors

Vectors respond differently to the action of pressing ENTER, DELETE, or BACKSPACE. The differences enhance the entry and editing of vectors. You can add or subtract elements to a vector, and you can combine two adjacent cells in a vector.

### ▶ To add an element to a vector by splitting an existing element

1. Place the insertion point in the cell that is to be split.

2. Press ENTER.

### ▶ To combine two adjacent cells in a vector

1. Place the insertion point next to the cell boundary you want to delete.

2. Press DELETE if the insertion point is to the left of the cell boundary or BACKSPACE if it is to the right of the cell boundary.

## Entering Case Expressions

A case expression is of this form

$$f(x) = \begin{cases} 0 \text{ if } x < 0 \\ 1 \text{ if } x \geq 0 \end{cases}$$

The right side of the equation is a matrix surrounded by brackets. The right bracket is an empty bracket.

▶ **To enter a case expression**

1. Enter the expression to the left of the $=$.

2. Enter the $=$.

3. Create a matrix.

   In most cases, the matrix should have three columns and at least two rows.

4. Enclose the matrix in brackets with an empty right bracket.

   The empty bracket appears on the screen as a dotted line, but it does not print.

5. Enter the case expression.

▶ **To align a case expression**

• Edit the properties of the matrix to align the expression as you want.

## Entering and Editing Multiline Displays

In *SW*, you can create multiline displays. Use multiline displays instead of a series of single line displays when you want to align each of the lines. For example, here are two separate displays:

$$x = a + \cos\theta - b$$
$$y = b + \sin\theta$$

These two equations look much better if they are aligned on the $=$ signs using a multiline display:

$$
\begin{aligned}
x &= a + \cos\theta - b\\
y &= b + \sin\theta
\end{aligned}
$$

Each line of a multiline display can have an optional label that becomes an automatically generated number when the document is printed. You can add a key to the label to create a cross-reference to the line in the document. See "Creating In-line and Displayed Mathematics" earlier in this chapter for more information about keys. By default, *SW* doesn't number equations when you preview or print your document, but you can change the default from the **User Setup** dialog box. See Chapter 13 "Customizing SW" for more information.

▶ **To create a multiline display**

1. Enter a display.

2. Type the first line and press ENTER to start the next line.

*SW* automatically aligns the lines on the first binary relation ($=$, $<$, etc.) in each line. You can choose your own alignment point if you want to override this automatic behavior.

### Aligning Multiline Displays

*SW* automatically aligns multiline displays according to the standard rules of LaTeX. If you prefer, you can change the alignment of a display line by selecting those characters on which you want LaTeX to align the expression. Although the style you choose for your document determines how multiline displays appear in print, you can add additional space between the lines.

If the style of your document uses the AMS package, you can center each line of the display or you can specify that the contents of the display be aligned as a single equation on multiple lines. For more information about AMS packages, see Chapter 10 "Applying Styles to Documents."

▶ **To set the alignment of a line in a display**

1. Select the characters on which you want to align the expression.

   –or–

   Place the insertion point so that it is at the left of the character on which you want to align the expression.

2. From the Edit menu, choose Set Alignment (ALT+E, A).

   The program inserts a special alignment character at the insertion point. This character is invisible unless you have turned on Invisibles from the View menu.

▶ **To center each line of a display**

1. Select the display.

2. Choose Properties.

3. Choose Advanced.

4. Check Enable Alignment.

5. Check Each Line Centered.

6. Choose OK.

▶ **To align the display as a single equation on multiple lines**

1. Select the display.

2. Choose Properties.

3. Choose Advanced.

4. Check Enable Alignment.

5. Check Single Equation on Multiple Lines.

6. Choose OK.

► **To add space after a line**

1. Select the display.

2. Choose **Properties**.

3. In the **Line Number** box, select the line after which you want to add space.

4. Specify a value for the amount of space you want to add, and indicate the unit of measure for the value.

5. Choose **OK**.

## Labeling Multiline Displays

You can edit the properties of a multiline display to add optional labels for each line of the display and for the display as a whole. When the document is printed, the label becomes an automatically generated equation number. You can override the automatically chosen number and supply your own number. Alternatively, you can use a single number for the entire display and add a letter to each line to indicate its sequence within the display. The sequence can continue to subsequent lines in the display, even if they are separated by text from the first part of the display.

As with a single-line display, you can supply an optional key for each line or, if you letter each line, for the entire display, then use the key to create cross-references to the line or equation from elsewhere in the text. When the document is printed, the reference is replaced by the number assigned to the line.

► **To number a line of a multiline display**

1. Place the insertion point inside the display at the end of the line you want to label.

   You can place the insertion point at any position that is not inside a template—the end of the line is a convenient position.

2. Press TAB.

   The program displays a gray box with a # to indicate that a number will be generated when you print or preview your document. The form of this number is determined by the typesetting style you choose for the document. Custom numbers and keys, if assigned, are shown in the box.

► **To enable lettering for a multiline display number**

1. Select the display.

2. Choose **Properties**.

3. Choose **Advanced**.

4. Check **Enable Lettering**.

5. To continue the lettering sequence from the previous display, check **Lettering Continuation**.

6. Choose **OK**.

▶ **To supply your own equation number**

1. Select the display.

2. Choose **Properties**.

3. Select the line you want to number.

4. Choose **Custom** in the **Number** box.

   The program displays a box for you to enter the number of the equation.

5. Type your number in the **Custom Number** box.

6. If you are using an AMS style that calls for equation number enclosures and you want to remove the enclosure, check **Suppress Annotation**.

7. Choose **OK**.

▶ **To add a key to a numbered line**

1. Select the display.

2. Choose **Properties**.

3. Select the number of the line you want.

4. In the key box, type a key for this line of the display.

5. Choose **OK**.

▶ **To add a key for the whole multiline display**

1. Select the display.

2. Choose **Properties**.

3. Choose **Advanced**.

4. Choose **Enable Lettering**.

5. In the **Key for the Whole Display** box, enter a key.

6. Choose **OK**.

## Entering Mathematical Names

*SW* can automatically recognize a special set of multicharacter names, including many common functions, as you type them. The recognized math names are listed in the **Automatic Substitution** dialog box. By default, *SW* displays recognized math names on the screen in gray upright text. For example, if you type **cos** in mathematics, the *co* appears in italics until you type the *s*, then the entire name changes to upright and is grayed. To have *SW* recognize math names automatically, the insertion point must be in mathematics and the **Disable automatic substitution** checkbox in the **Automatic Substitution** dialog box must be unchecked.

### Entering Math Names

▶ **To enter a recognized math name**

- On the Math toolbar, click the Math Name button  **sin cos** .
  –or–

- From the **Insert** menu, choose **Math Name** (ALT+I, N).
  –or–

1. Make sure the **Disable Automatic Substitution** checkbox in the **Automatic Substitution** dialog box is unchecked.

2. Start mathematics.

3. Type the name.

If you type the math name, *SW* inserts the name directly into your document. If you use either of the last two methods, *SW* displays the **Math Name** dialog box so you can select the name you want.

---

**Tip** If you want to enter $\cos h$, $\sin h$, or $\tan h$ instead of $\cosh$, $\sinh$, or $\tanh$, type the first three letters, press SPACEBAR, and then type **h**. Pressing the SPACEBAR tells *SW* to stop searching for additional characters to add to the function name.

---

You can also create multicharacter variable names for use in mathematics. For example, you can create the names Force, Mass, and Acceleration, and assign a value to Mass and Acceleration. Then, you can write the equation

$$\text{Force} = \text{Mass} \times \text{Acceleration}$$

and compute the value of Force. If you start mathematics and enter the words Force, Mass, and Acceleration without defining them as multicharacter variable names, the program considers each letter in each word to be a variable. *SW* can automatically recognize custom math names. See Chapter 13 "Customizing SW" for more information.

▶ **To create a custom math name**

1. Start mathematics.

2. On the Math toolbar, click the Math Name button  .

   –or–

   From the Insert menu, choose **Math Name** (ALT+I, N).

3. In the **Name** box, type the name you want to create.

4. Check the button to indicate the nature of the math name.

5. If the math name is an operator, click the button that determines where you want the limits to be placed.

6. Choose **OK**.

### Adding Limits to Math Names

Some math names, such as max and lim, can have limits. When a math name is in-line, the program places the limits at the right:

$$f(x) = \lim_{n \to \infty} x_n$$

When the math name is displayed, the program places the limits above or below:

$$f(x) = \lim_{n \to \infty} x_n$$

You can override these defaults in the **Math Name** dialog box when you enter a math name or by editing the properties of an existing math name.

▶ **To add limits to a math name**

1. Enter the math name.

2. Use the **Subscript** and **Superscript** commands to enter the limits.

   You can create math names with multiline limits. See "Entering Operators" earlier in this chapter for details.

## Entering Horizontal Spaces in Mathematics

$SW$ automatically provides appropriate spacing for mathematics. In some situations, however, $SW$ does not have enough information to space your mathematics exactly as you want them. For example, in the expression

$$\int_a^b f(x)\, dx$$

a little extra space (a thin space) is required between $f(x)$ and $dx$. You can add these horizontal spaces inside mathematical expressions:

| Name | Size |
|------|------|
| Em space | width of M |
| 2-em space | width of MM |
| Normal | $\frac{1}{6}$ em |
| Required | $\frac{1}{6}$ em |
| Nonbreaking | $\frac{1}{6}$ em |
| Thin space | $\frac{2}{9}$ em |
| Thick space | $\frac{5}{18}$ em |
| Zero space | 0 em |
| Negative thin space | $-\frac{2}{9}$ em |
| Italic correction | Depends on character to left |

Some horizontal spaces have keyboard shortcuts. See Appendix B "Keyboard Shortcuts and Key Prefixes" for more information.

The zero space, which translates to the TEX empty group, is used with prescripts such as the 2 in this expression

$$_2Z_3$$

To enter the expression $_2Z_3$, enter a zero space, enter 2 as a subscript, then complete the expression. Since subscripts and superscripts adjust to the size of the objects to their left, placing them after a zero space ensures that they appear at a fixed position.

#### ▶ To enter spaces

1. On the Math toolbar, click the Spaces button .

    –or–

    From the Insert menu, choose Spacing, and then choose Horizontal Space (ALT+I, S, H).

    Both methods open the Horizontal Space dialog box.

2. Check the button for the space you want.

3. Choose OK.

## Entering Labels

In *SW*, label templates have two parts: the expression to be labeled and the label itself. The label is set in the subscript/superscript type style. In the formula

$$\underbrace{x + \cdots + x}_{k \text{ times}}$$

the label is "$k$ times" and the expression being labeled is $\underbrace{x + \cdots + x}$.

▶ **To enter a label**

- On the Math toolbar, click the Label button .
  −or−
- From the Insert menu, choose Label (ALT+I, L).

  Both methods open the Label dialog box from which you can choose the template you want.

▶ **To apply a label to an existing expression**

1. Select the expression.

2. Enter a label.

▶ **To remove a label from an expression**

1. Place the insertion point to the left of the entire labeled expression.

2. Press DELETE.

   −or−

1. Place the insertion point to the right of the entire labeled expression.

2. Press BACKSPACE.

▶ **To move the label to the top or bottom**

1. Select the entire labeled expression.

2. Choose Properties.

3. From the Label Properties dialog box, choose the label position you want.

4. Choose OK.

## Entering Decorations

*SW* provides templates for placing horizontal bars, braces, or arrows above or below an expression:

$$\overline{x+y} \qquad \overleftarrow{x+y} \qquad \overrightarrow{x+y} \qquad \overleftrightarrow{x+y} \qquad \overbrace{x+y}$$

$$\underline{x+y} \qquad \underleftarrow{x+y} \qquad \underrightarrow{x+y} \qquad \underleftrightarrow{x+y} \qquad \underbrace{x+y}$$

In addition, the program has two special wide accents:

$$\widehat{xyz} \qquad \widetilde{xyz}$$

You can place a box around an expression. The box can be exactly the size of the expression or can be set off from the expression on all sides with a little extra space:

$$\boxed{x+y=z} \qquad \boxed{\,x+y=z\,}$$

Collectively, these bars, arrows, wide accents, and boxes are known as *decorations*.

### ▶ To enter a decoration

- On the Math toolbar, click the Decoration button $\boxed{\leftrightarrow}$ .
  —or—
- From the Insert menu, choose Decoration (ALT+I, C).

Both methods open the Decoration dialog box from which you can choose the effect you want.

### ▶ To apply a decoration to an existing expression

1. Select the expression.

2. Enter a decoration.

### ▶ To remove the decoration from an expression

1. Place the insertion point to the left of the template.

2. Press DELETE.

   —or—

1. Place the insertion point to the right of the template.

2. Press BACKSPACE.

### ▶ To change the decoration

1. Select the entire template or place the insertion point to its right.

2. Choose Properties.

3. From the Decoration Properties dialog box, choose the decoration you want.

4. Choose OK.

### Entering Primes

The $SW$ prime symbol is a large character that sits on the baseline. You can create expressions such as $y'$ by placing the prime symbol in a superscript. The superscript raises the character and scales down its size.

You can create expressions containing primes either with the mouse or with the keyboard, but the two methods are very different. If you use the mouse to enter a prime, $SW$ neither positions nor scales the symbol. If you use the keyboard to enter a prime, $SW$ positions and scales the symbol automatically.

▶ **To enter a prime with the mouse**

1. Insert a superscript template.

2. Enter the prime character from the Miscellaneous Symbols panel on the Symbol toolbar.

3. Return the insertion point to the main line.

▶ **To enter a prime with the keyboard**

• Press the right quotation mark key (') when the insertion point is in mathematics. The program inserts the prime character in a superscript and then returns the insertion point to the main line.

▶ **To add a prime to a character that already has a superscript**

1. Place the insertion point in the superscript where you want the prime.

2. Select the prime from the Miscellaneous Symbols panel on the Symbol toolbar.

### Entering Degrees

You enter a degree symbol in expressions such as $\sin 30°$ by inserting a small circle symbol in a superscript.

▶ **To enter a degree symbol**

1. Insert a superscript template.

2. Enter the small circle character from the Common Symbols toolbar.

    –or–

    Type CTRL+**circ.**

3. Return the insertion point to the main line.

# 5 Previewing and Printing Documents

*SW* previews and prints your documents in several steps:

1. The program compiles your document with LaTeX to create a device-independent, or DVI, file.

   The compilation resolves any internal cross-references and creates any required document parts that must be generated, such as a table of contents, an index, or a bibliography. The complexity of your document determines how many times *SW* passes it through LaTeX. The DVI file that results from the compilation contains your typeset document in a form independent of any output device.

2. The program sends the DVI file to the TrueTeX screen previewer, which displays your document as it will appear in print, or to the printer driver, which prints your document on the printer or to a file.

   You can complete both steps at the same time as part of the preview or print process, or you can compile your document at one time and preview or print it later.

   The LaTeX compiler, the screen previewer, and the printer driver are all supplied with *SW*. See "How SW Uses DVI Files" later in this chapter for more information about how the program works with compiled documents.

## Previewing a Document

Previewing gives you an opportunity to see how your document will appear in print and to check for any changes you need to make. Careful previewing can save a considerable amount of time and paper.

When you preview, *SW* compiles the document with LaTeX if a valid DVI file for the document is not already present. If your document has cross-references, two LaTeX passes are necessary. If you have requested a table of contents, three LaTeX passes are necessary. The LaTeX dialog box indicates which pass is in progress. While LaTeX is compiling a file, the LaTeX icon—a lion in sunglasses—appears in the lower-left corner of the main window. *SW* is suspended until the DVI file is written.

After the compilation, *SW* sends the DVI file to the TrueTEX screen previewer, which displays your typeset document on the screen exactly as it will appear in print.

---

**Note**   If you save a document immediately before you preview it, *SW* retains the DVI file it creates during the preview process. If the document doesn't change before the next time you preview or print it, *SW* skips the compilation step and sends the DVI file directly to the previewer or the printer driver. If you don't save the document immediately before you preview it, *SW* recompiles the document the next time you preview or print, even if it hasn't changed.

---

### ▶ To preview a document

1. Open the document you want to preview.

2. On the Standard toolbar, click the Preview button 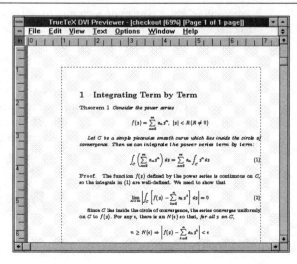 .

   –or–

   From the File menu, choose **Preview** (ALT+F, W).

   The program looks for a valid DVI file for your document.

   - If it finds a file, the program displays the DVI file in the preview screen.
   - If it doesn't find a file, the program compiles your document with LaTEX. While it is compiling, the program displays the LaTeX dialog box, which indicates the number of times your document will be passed through LaTEX and the number of the pass in progress. When LaTEX has created a DVI file, the previewer displays your document as it will appear in print.

---

**Note**   See "Compiling a Document" later in this chapter for information about working with LaTEX and correcting any LaTEX errors that might occur during compilation.

---

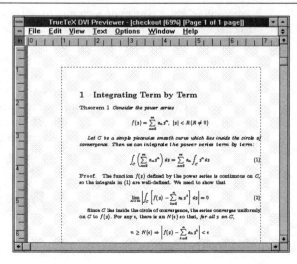

The TrueT<sub>E</sub>X preview screen has scroll bars at the bottom and the right and a menu bar across the top that you can use to move around in your document and to customize the way the previewed document appears on the screen. Instructions for using other options appear in online Help for the TrueT<sub>E</sub>X Previewer.

▶ **To move around on a previewed page**

- Use the scroll bars at the right and bottom of the screen.
  −or−
- Use the ARROW keys.
  −or−
- Use the commands on the View menu.

▶ **To move to another page in the previewed document**

- Press PAGE UP or PAGE DOWN.
  −or−
- From the Edit menu, choose Go to Page and specify the page you want to see.
  −or−
- Use the commands on the View menu.

▶ **To magnify the size of the display**

- Drag the mouse to draw a box around the portion of text you want to magnify.
  −or−
- From the View menu, choose Zoom in.

▶ **To reduce the size of the display**

- From the View menu, choose Zoom out.

▶ **To leave the preview screen**

- From the File menu, choose Exit (ALT+F, X).
  The previewer returns you to your document, where you can make any necessary changes.

## Printing a Document

You can print a document from the preview screen or from the *SW* main window.

### Printing from the Preview Screen

If your document needs no further changes when you preview it, you can print it directly from the preview screen. You may need to select a printer. You can also print the document to a file.

► **To select a printer from the preview screen**

1. From the File menu, choose Print Setup (ALT+F, R).

2. In the Name box, enter the name of the printer you want to use.

3. In the Paper Size box, enter the size of the paper in the printer.

4. In the Paper Source box, enter the paper source for the printer.

5. If you need to change the paper orientation, click the button corresponding to the orientation you want.

6. Choose OK.

► **To print the document from the preview screen**

1. From the File menu in the previewer, choose Print (ALT+F, P).

2. Select the printer you want.

3. Make these selections in the Print dialog box:

   - If you need to change the orientation or size of the paper in the printer, choose Properties, make the changes you want, and then choose OK.
   - Set the range of pages to be printed. The default is all pages.
   - Set the number of copies to be printed. The default is one copy.
   - Check Collate if you want the copies collated.

4. When all the settings are correct, choose OK.

The previewer displays a print status bar at the top of the document image as it sends your document to the printer. The bar indicates which page is being sent.

► **To print to a file**

1. From the File menu in the previewer, choose Print (ALT+F, P).

2. In the Print dialog box, select a print device that is set up to print to a file.

   If no print device is set up to print to a file, see your Windows documentation for instructions.

3. Check the Print to file box.

4. Set the range of pages to be printed.

5. Choose OK.

The print status bar at the top of the document image indicates which page is being printed to the file.

▶ **To cancel printing**

- Click the print status bar.

## Printing from the SW Main Window

When you print, *SW* compiles the document if a valid DVI file for the document is not already present. While LaTeX is compiling a file, the LaTeX icon appears in the lower-left corner of the main window. *SW* is suspended until the DVI file is written. Once the document has been compiled, *SW* sends it through the printer driver to a printer or to a file.

Remember that you can save a considerable amount of time and paper if you preview your document carefully before you print it.

▶ **To send a document to the printer**

1. Open the file you want to print.

2. On the Standard toolbar, click the Print button ⊞ .

   –or–

   From the File menu, choose Print (ALT+F, P).

3. Make these selections in the Print dialog box:

   - Select the printer
     *SW* prints to any Windows print device, including facsimile devices. If you need to install a print device, refer to your Windows documentation for instructions.
   - Select the pages you want to print. The default is all pages.
   - In the Step box, set a step factor for printing. The default is 1, or every page. To print all odd pages, select 1 as the first page and step by 2. To print all even pages, select 2 as the first page and step by 2.
   - Select the number of copies you want to print. The default is one copy.
   - If you want to print your document from last page to the first, check Print in Reverse Order.

4. When all the settings are correct, choose OK.

   The program checks to see if a valid DVI file is present for your document.

   - If it finds a file, it bypasses LaTeX and prints the DVI file.
   - If it doesn't find a file, it compiles your document with LaTeX.

   The program automatically sets the number of LaTeX passes, depending on the complexity of your document. If your document contains cross-references, the number of passes should be 2. If you have requested a table of contents, the number of passes should be 3. When LaTeX has created a DVI file, *SW* sends the file to the printer.

   See "Compiling a Document" later in this chapter for information about working with LaTeX and correcting any LaTeX errors that might occur during compilation.

## Compiling a Document

You can compile your document independently of the preview or print process. When you compile, *SW* sends the document to LaTeX to create a DVI file. While LaTeX is compiling the file, the LaTeX icon appears in the lower-left corner of the main window. *SW* is suspended until the DVI file is written.

To compile the document, use the Compile command on the File menu. The Compile command is active only when you have saved the document and made no further changes.

▶ **To compile a document independently of the preview or print process**

1. Save your document.

2. From the File menu, choose Compile (ALT+F, L).

   The program displays the Compile dialog box.

3. Select the options you want:

   - If your document contains a BibTeX bibliography field and you want to create the bibliography, check Generate a Bibliography.
   - If your document contains index entries and you want to create the index, check Generate an Index.

4. Choose OK.

   The program displays the LaTeX dialog box, which indicates the number of times your document will be passed through LaTeX and the number of the pass in progress. When LaTeX has created a DVI file, the program returns to the main screen.

## Working with LaTeX

When you compile, preview, or print a document, LaTeX resolves the cross-references internal to your document and creates a DVI file. The legend under the LaTeX icon summarizes the DVI file-creation process and indicates which page of your document is being compiled. You can view the LaTeX process in detail by opening the LaTeX window.

▶ **To open the LaTeX window**

- Double-click the LaTeX icon when it appears during compilation.

▶ **To close the LaTeX window and let the compilation continue**

- Click the down arrow in the upper-right corner of the LaTeX window.

▶ **To stop the current LaTeX pass**

• From the File menu in the LaTeX window, choose Exit (ALT+F, X).
  –or–

• Double-click the Control box in the upper-left corner of the LaTeX dialog box.
  –or–

• If the LaTeX window is open, press ALT+F4.

## Using External Processing

If your computer is connected to a network server that can run LaTeX, you may be able to set up *SW* so that document files can be sent directly to this server for LaTeX compiling and printing. If you can use the server, you can avoid waiting for LaTeX to compile your file. The Technical Reference, which is supplied in the *SW* document `techref.tex` in the `extras` directory, contains further information about external processing.

## Correcting LaTeX Errors

If LaTeX is unable to compile your document, the LaTeX window opens and displays an error message and waits for user input. The message will help you identify and correct the error. The most common cause of problems is the entry of incorrect TeX codes in a TeX field.

▶ **To identify a LaTeX error**

1. Find the error message in the LaTeX window.

2. Look for a line number.

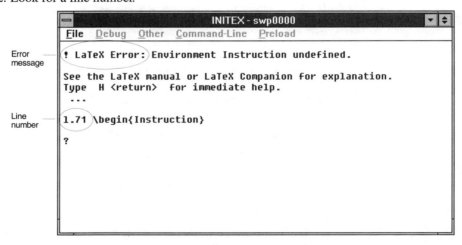

The line number indicates where the error has occurred. It usually refers to a line in your document, although it can also refer to a line in one of the included style files.

3. Close the LaTeX window.

Since lines in an *SW* document do not correspond to lines in the .tex file, you may need to use an ASCII editor to locate the error. A familiarity with TeX and LaTeX will help you find and correct the problem.

LaTeX also records error messages in a file with the .log extension. If you saved your document before printing or previewing, the .log file is placed in the same directory as the .tex file. If you haven't saved your document before printing or previewing or if your document is located on a remote or read-only drive, *SW* doesn't write to it. Instead, it copies your document to your local temp directory, along with the .log file. If you request technical assistance from TCI, you may be asked for a copy of the .tex file and the .log file.

## How SW Uses DVI Files

*SW* maintains DVI files that correspond to the saved versions of documents. In other words, if you save your document before you compile, preview, or print it, *SW* maintains the DVI file for the document. If you don't change the document before you preview or print it again, *SW* skips the LaTeX compilation and sends the DVI file directly to the previewer or the printer driver.

If, on the other hand, you make changes to your document but don't save it before you preview or print, *SW* discards the new DVI file after the preview or print is complete. The next time you preview or print, the document must be recompiled.

## Using the TeX Icons Directly

If your document has correct LaTeX syntax, you can use the icons in the *SW* program group to compile, preview, and print your document. Choose the TrueTeX Formatter icon to compile the file and the TrueTeX Previewer icon to preview or print it. If your document is a subdocument (see Chapter 12 "Managing Documents"), you must compile, preview, and print it from within *SW*.

▶ **To compile a document from outside SW**

1. Double-click the TrueTeX Formatter icon in the *SW* program group.

   The TrueTeX Formatter window opens and within it, the **Open TeX File** dialog box.

2. Select the document file and directory.

3. Click **OK**.

   LaTeX creates the DVI file and places it in the same directory as the document file. The two files will have the same file name but different file extensions: .tex for the document file and .dvi for the DVI file.

▶ **To preview a document from outside SW**

1. Double-click the TrueTEX Previewer icon in the *SW* program group to open the preview screen.

2. From the File menu, choose Open (ALT+F, O).

   The program opens the Open DVI File dialog box, which lists all .dvi files in the current directory.

3. Select the DVI file for the document you want to preview.

4. Click OK.

   The previewer displays the DVI file in the preview screen.

▶ **To print a document from outside SW**

1. Click the TrueTEX Previewer icon in the *SW* program group to preview the document.

2. From the File menu, choose Print (ALT+F, P).

3. Make the selections you want in the Print dialog box.

   For more information, see "Printing from the Preview Screen" earlier in this chapter and the Help menu in the previewer.

4. Choose OK.

# 6 Using Special Features

Special *SW* features help you enhance your documents with lists, fragments, cross-references, hypertext links, and external program calls.

## Creating Lists

An *SW* list is a series of items, each consisting of one or more paragraphs. You create lists by tagging paragraphs as *list items*. The item tags for creating lists appear in the Item Tag popup list on the Tag toolbar.

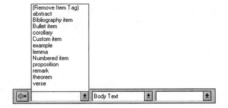

The list of item tags you see depends on the typesetting style you have chosen for your document, but almost all styles support these standard LaTeX list items: bulleted items, numbered items, and custom items. *SW* supports lists up to four levels deep. The item tag in effect at the insertion point is named in the Item Tag popup box. The Remove Item Tag button at the left of the Tag toolbar removes the most recently applied item tag in the paragraph containing the insertion point.

Each list item contains a *lead-in*—a bullet, a number, or some customized text you supply—and the item paragraph. To create a list, you enter the lead-in you want, then enter the paragraph, or you apply the item tag for the kind of list you want to several existing paragraphs. You can also create nested lists. The sections below explain how to create lead-ins; lists with bulleted, numbered, or custom item tags; and nested lists.

---

**Note** Bibliographies are lists of references. You can create a bibliography manually with the Bibliography item tag, or you can use the BIBTeX feature. Refer to Chapter 11 "Structuring Documents" for more information on creating bibliographies.

---

▶ **To enter a lead-in**

- Use the Item Tag popup list:

    a. Click the Item Tag popup list on the Tag toolbar.
    b. Click the item tag for the kind of list you want.

    –or–

- Use the **Apply** command:

  a. From the **Tag** menu, choose **Apply** (ALT+T, A).
  b. Select the item tag for the kind of list you want.
  c. Choose **OK**.

  –or–

- Press the function key assigned to the item tag you want.
  When you first install *SW*, function key assignments are in effect for the Numbered item tag (F7) and the Bullet item tag (F8). Refer to Chapter 13 "Customizing SW" for information about changing function key assignments.

  *SW* inserts the appropriate lead-in and indention for the first list item. Each time you begin a new paragraph, the program assumes you're starting a new list item and automatically provides the correct lead-in and indention. If you want the new paragraph to be part of the preceding list item, press BACKSPACE to delete the lead-in and then type the paragraph. Lead-ins not followed by text, or empty list items, are deleted by the program when you reopen your document, unless the lead-in has been customized from the **Lead Item Properties** dialog box.

### ▶ To end a list

1. When you have entered all the list items you want, press ENTER.

2. Click the Remove Item Tag button .

   –or–

   Press F2 to end the list.

   The program removes the lead-in and returns to the level of indention before the list.

### ▶ To make existing paragraphs into list items

1. Select the paragraphs.

2. Apply the item tag for the type of list you want.

### ▶ To open the Lead Item Properties dialog box

- Double-click the lead-in.
  –or–

1. Place the insertion point to the right of the lead-in box.

2. Choose Properties:

   - On the Standard toolbar, click the Properties button 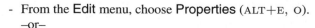.
     –or–
   - From the **Edit** menu, choose **Properties** (ALT+E, O).
     –or–
   - Press CTRL+F5.

## Bulleted Lists

Lead-ins for bulleted lists typically contain bullets at the first level and other symbols at lower levels, as in this list:

- Level 1
  - \* Level 2
    - − Level 3

On the screen, a bulleted lead-in appears as a small yellow box containing a bullet. When you preview or print, *SW* replaces the lead-in by the bullet symbol. The symbols used at each level in the printed document are determined by the typesetting style you choose for the document.

### ▶ To create a bulleted list

1. Press ENTER to start the first paragraph of the first list item.

2. Choose the Bullet item tag.

   The program automatically inserts a lead-in that is replaced by a bullet when the document is printed. On the screen, the lead-in appears as a small yellow box containing a bullet.

3. Type the text of the list item.

4. Press ENTER to end the item.

   If the item contains more than one paragraph, press BACKSPACE to delete the automatic lead-in at the beginning of each continuation paragraph.

5. Enter each list item as you did the first, pressing ENTER at the end of each item.

6. To end the list, click the Remove Item Tag button .

   −or−

   Press F2.

   The program removes the lead-in and returns to the level of indention before the list.

## Numbered Lists

Lead-ins in numbered lists may contain uppercase or lowercase Roman numerals, Arabic numerals, or uppercase or lowercase alphabetic characters, depending on the typesetting style you choose. Usually, the lead-ins differ at each level, as in this list:

1. Level 1

   a. Level 2

      i Level 3

On the screen, a numbered lead-in appears as a small yellow box containing a number sign (#). When you preview or print your document, *SW* replaces the lead-in by an automatically generated number. You can restart the automatic numbering of the list at any point.

### ▶ To create a numbered list

1. Press ENTER to start the first paragraph of the first list item.

2. Choose the Numbered item tag.

    The program inserts the lead-in.

3. Type the text of the list item.

4. Press ENTER to end the item.

    If the item contains more than one paragraph, press BACKSPACE to delete the automatic lead-in at the beginning of each continuation paragraph.

5. Enter each list item as you did the first, pressing ENTER at the end of each item.

6. To end the list, click the Remove Item Tag button .

    –or–

    Press F2.

    The program removes the lead-in and returns to the level of indention before the list.

### ▶ To restart the lead-in numbering

1. Open the **Lead Item Properties** dialog box:

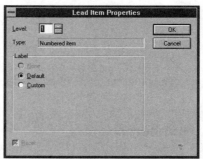

    If you have a nested list, the lead-ins are listed with the outermost list first.

2. In the **Level** box, set the level of the numbered lead-in you want to reset.

    The program will set the numbering back to 1 at this lead-in.

3. Check **Reset**.

4. Choose **OK**.

## Custom Lists

The lead-ins for custom lists contain text that you supply. The typesetting style you choose for your document determines the appearance of the lead-in but not its content. A lead-in is limited to a single paragraph. Its length determines the left edge of the text when you print the document. You can enter text, mathematics, and special characters in the lead-in and you can apply tags; for example,

One $\left(a^2\right)$     First item.

**Two** $\left(b^2\right)$     Second item. The lead-in has bold text.

Three $\left(c^2\right)$     Third item.

▶ **To create a custom list**

1. Press ENTER to start the first paragraph of the first list item.

2. Choose the Custom item tag.

    The program automatically inserts the lead-in, which appears on the screen as a small yellow box.

3. Open the **Lead Item Properties** dialog box:

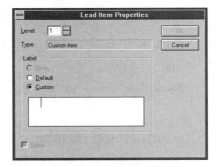

4. Choose **Custom**.

5. In the text box, type the text you want to appear in the lead-in.

6. Choose **OK**.

    Your text appears in the yellow box, like this: Your Text.

7. Type the text of the list item.

8. Press ENTER to end the item.

    If the item contains more than one paragraph, press BACKSPACE to delete the automatic lead-in at the beginning of each continuation paragraph.

9. Enter each list item as you did the first, pressing ENTER at the end of each item.

10. To end the list, click the Remove Item Tag button .

–or–

Press F2.

The program removes the lead-in and returns to the level of indention before the list.

## Nested Lists

You can nest one list within another by applying additional tags to a paragraph. You can apply up to four tags to a paragraph and therefore nest items four deep. Although the Tag toolbar indicates only the latest item tag applied to a paragraph, the on-screen indention associated with each tag shows you when a paragraph has more than one tag.

▶ **To create a nested list**

• Start a new list while inside an existing list.

The program creates and displays a lead-in item for the nested list.

▶ **To leave an inner list and return to the next outer list**

1. Press ENTER to create a new paragraph.

2. Click the Remove Item Tag button .

–or–

Press F2 to remove the innermost list item tag.

## Customizing a Numbered or Bulleted List

Although *SW* automatically provides the lead-in for all numbered or bulleted lists, you can customize or remove the lead-in. Customizing a numbered or bulleted lead-in has a different effect from creating a custom list. In a numbered or bulleted list, the text aligns at the left, even if you customize the lead-in. Therefore, a customized lead-in for a numbered or bulleted item may extend into the left margin of the page, like this:

One. The first item in a customized numbered list.

Two. The second item.

Three. The third item.

In a custom list, however, the length of the lead-in determines the left edge of the text:

One.      The first item in a custom list.
Two.      The second item.
Three.      The third item.

▶ **To customize a numbered or bulleted list lead-in**

1. Select the lead-in you want to customize.

2. Open the **Lead Item Properties** dialog box.

3. Customize the lead-in:

    a. In the **Level** box, set the level of the lead-in you want to customize.
    b. Choose **Custom**.
    c. Type the lead-in.
       In the lead-in, you can use text, mathematics, and special characters, and you can apply tags to the text. $SW$ replaces the default lead-in with the lead-in you specify. If you provide a custom lead-in for a numbered item, the program suspends automatic numbering for the item.

4. If you want to remove the lead-in, choose **None**.

5. Choose **OK**.

# Using Fragments

A *fragment* is a portion of a paragraph you save in a separate file for later recall. You can use fragments for saving equations or phrases that occur frequently in your work. A fragment saved in one document is available to all documents.

When you recall a fragment, its content is pasted into your document at the insertion point. If a fragment contains a field, the field appears as a small gray box on the screen. When you print the document, the field is interpreted and printed correctly.

## Predefined Fragments

$SW$ comes with several fragments already defined and stored. Some predefined fragments contain equations or theorems common in mathematics, such as the ones that follow.

- quadrat (the roots of the quadratic equation):
$$\frac{-b \pm \sqrt{b^2 - 4ac}}{2a}$$

- erf (the error function):
$$\int_{-\infty}^{\infty} e^{-x^2} \, dx = \sqrt{\pi}$$

- meanval (a statement of the mean value theorem):

    Let $f$ be a continuous function on the closed interval $[a, b]$ and have a derivative at every $x$ in the open interval $(a, b)$. Then there is at least one number $c$ in the open interval $(a, b)$ such that $f'(c) = \frac{f(b)-f(a)}{b-a}$.

## User-Defined Fragments

You can save as fragments those portions of text and mathematics you use most often. Fragments can contain both text and mathematics; they must be contained in a single paragraph. The fragments you define are available to all documents.

▶ **To define a fragment**

1. Open the document that contains the information you want to save as a fragment.

   —or—

   Open a file and type the information.

2. Select the part of the document you want to save as a fragment.

   The selection must lie entirely within a single paragraph.

3. From the File menu, choose Save Fragment (ALT+F, G).

4. Type a file name to be used to recall the fragment.

   The name must be eight characters or less and must use only DOS file name characters. See Chapter 2 "Opening and Closing Documents" for information about DOS file names. Avoid using the name of a TEX command for your fragment. See Appendix B "Keyboard Shortcuts and Key Prefixes" for a list of TEX commands recognized by *SW*.

5. Choose OK.

   The program saves your fragment and immediately inserts its name in the Fragments popup list on the Fragments toolbar.

## Inserting Fragments

You can insert a fragment into your document in several ways.

▶ **To paste a fragment at the insertion point**

1. Click the Fragments popup list.

2. Click the name of the fragment you want.

   —or—

1. From the File menu, choose Import Fragment (ALT+F, R).

2. From the File Name area in the Import Fragment dialog box, select the fragment you want.

3. Choose OK.

–or–

- Press CTRL+***name***, where ***name*** is the name of the fragment you want to insert. This method works only for fragments with names at least two characters long.

---

**Note** If you use CTRL+***name*** to insert a fragment that has the same name as a TEX command, *SW* inserts the TEX command instead of the fragment. See Appendix B "Keyboard Shortcuts and Key Prefixes" for a list of TEX commands that *SW* recognizes.

---

## Deleting Fragments

You can remove a fragment from the Fragments popup list.

▶ **To remove a fragment from the Fragments popup list**

- Delete the file with the fragment name and the extension .frg from the fragments directory.

# Creating Notes

Paragraphs can contain references to notes that appear elsewhere in the document, such as footnotes like this[1] or margin notes like the one that appears on this page. Notes can contain both text and mathematics.

    *SW* automatically numbers footnotes sequentially throughout your document unless you override the numbering. On the screen, the footnote appears as a small gray box containing the word *footnote,* like this:  footnote . When you print the document, *SW* prints the footnote number in the text and the footnote itself at the bottom of the page.

    Margin notes don't carry numbers. On the screen, margin notes appear as small gray

This is the text of a margin note.
boxes containing the words *margin note,* like this:  Margin note . When you print the document, *SW* places the note in the margin, beginning on the line in which you inserted the note.

▶ **To enter a footnote**

1. Position the insertion point where you want the footnote number to appear in the text.

2. On the Field toolbar, click the Note button  .

   –or–

   From the Insert menu, choose Field, and then choose Note (ALT+I, E, N).

   The program opens the Note dialog box with the Footnote box checked.

---

[1]    Use footnotes sparingly. Much of the information in footnotes is overlooked by the reader.

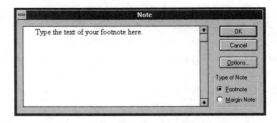

3. Enter the text of the footnote.

4. Choose **OK**.

The **Options** button in the **Note** dialog box provides a way for you to obtain non-standard footnotes. You can set up multiple references to a single footnote and you can add a footnote with no number or in-text reference.

▶ **To enter multiple references to the same footnote**

1. Enter the footnote at the first reference point.

2. At each subsequent reference point to the footnote, do the following:

    a. On the Field toolbar, click the Note button  .
       –or–
       From the **Insert** menu, choose **Field,** and then choose **Note** (ALT+I, E, N).
       Do not enter any footnote text.
    b. Choose **Options.**
    c. Check **Override Automatic Number.**
    d. Enter the number of the footnote in the **Footnote Number** box.
       You may need to preview the document to determine the correct number.
    e. Check **Mark Only.**
       When you choose **Mark Only,** the program places the footnote number in the text
       but doesn't print the footnote at the bottom of the page.
    f. Choose **OK** to leave the **Options** dialog box.
    g. Choose **OK** to return to your document.

▶ **To add a note at the bottom of the page with no number or in-text reference**

1. On the Field toolbar, click the Note button  .

   –or–

From the **Insert** menu, choose **Field,** and then choose **Note** (ALT+I, E, N).

2. Enter the text of the note.

3. Choose **Options.**

4. Check **Text Only.**

When you choose **Text Only**, the program doesn't place a footnote number in the text but prints the note at the bottom of the page.

5. Choose **OK** to leave the **Options** dialog box.

6. Choose **OK** to return to your document.

---

**Note**    In some environments, particularly within mathematics, footnotes are disallowed by LATEX. In those circumstances, you can create a footnote by placing a footnote marker in the place where the footnote is disallowed, then placing the text of the footnote in the next possible location where footnotes are allowed. Use the **Mark Only** option to create the marker and the **Text Only** option to create the text of the footnote.

---

▶ **To enter a margin note**

1. Position the insertion point in the line where you want the margin note to appear.

   If you want a note to appear close to a section heading, place the insertion point on the line following the heading.

2. On the Field toolbar, click the Note button  .

   –or–

   From the **Insert** menu, choose **Field**, and then choose **Note** (ALT+I, E, N).

3. Check **Margin Note**.

4. Enter the text of the margin note.

5. Choose **OK**.

# Creating Cross-References

*SW* paragraphs can also contain references to the numbers of other parts of the document, such as equations, graphics and tables, sections, bibliography items, and pages. Cross-references have two parts: a *marker* for the item, which contains a unique key, and the *reference* to the item, which uses the key to create the cross-reference. The marker links the key to the number of the document item (such as the section number or page number). When you preview or print a document containing cross-references, *SW* automatically replaces the key with the number of the document item.

---

**Note**    When you preview or print a document that uses cross-references, *SW* processes the document through LATEX twice to resolve the cross-references. See Chapter 5 "Previewing and Printing Documents" for more information about previewing and printing documents.

---

Cross-references behave like *hypertext links*. Once you have created a cross-reference, you can use it to move to the paragraph containing the referenced marker.

## Markers

To be able to refer in your text to sections, numbered list items, pages, equations, theorems, and tables and graphics, you must mark each one with a unique key. The marker links the key to the number of the document item, such as the section number or the number of an item in a numbered list.

To create markers for all items except mathematical displays and graphics, use the process described below. To create markers for numbered mathematical displays, use the **Properties** command, as described in Chapter 4 "Entering and Editing Mathematics." To create markers for graphics, use the **Frame** tab sheet in the **Graphics Properties** dialog box, as described in Chapter 7 "Using Tables and Graphics."

▶ **To create a marker**

1. Place the insertion point immediately after the item you want to mark.

   If you want to create a marker for a section heading, place the insertion point on the line following the heading.

2. On the Field toolbar, click the Marker button .

   –or–

   From the **Insert** menu, choose **Field**, and then choose **Marker** (ALT+I, E, M).

   The program opens the **Marker** dialog box.

3. Enter a unique key for the item.

   Use only letters, numbers, and the following characters in your keys: , : ; ? ! ' ' ( ) [ ] - / * . Spaces are acceptable.

4. If you want to display a list of the keys already defined for your document, click the arrow next to the **Key** box.

   Scroll the list to make sure you are not duplicating a key.

5. Choose **OK**.

   If the option to view marker fields is turned on, *SW* displays the marker on the screen in a small gray box containing the word *marker* and the key you entered for the item, like this:  marker: example . The marker does not appear in print.

▶ **To toggle the display of marker fields on and off**

- From the **View** menu, choose **Marker Fields** (ALT+V, K).

## References

You can create a reference to the number of any marked item, such as a section, page, equation, theorem, or graphic.

▶ **To create a reference to the number of an item**

1. Place the insertion point where you want the reference to appear.

2. On the Field toolbar, click the Cross Reference button 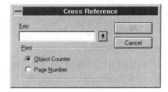 .

   –or–

   From the **Insert** menu, choose **Field,** and then choose **Cross Reference** (ALT+I, E, R).

   *SW* opens the **Cross Reference** dialog box.

3. Enter the key of the item.

   - If you want to create a reference to an item that already has a key

     a. Click the arrow next to the **Key** box to display the list of defined keys.

     b. Scroll the list to find the key you want.

     c. Click the name to select it.

   - If you want to create a reference to an item that doesn't yet have a key, enter the key in the **Key** box.

   You can define the reference before you define the key, but you must define both parts before the cross-reference is complete.

4. If you are creating a reference to a page, select **Page Number** (ALT+N).

5. If you are creating a reference to any other numbered item, click **Object Counter** (ALT+O).

6. Choose **OK**.

   *SW* displays the reference on the screen in a small gray box containing the word *ref* and the key you entered, like this:  ref: example . If you entered a page reference, the box contains the word *pageref* and looks like this:  pageref: example . When you print the document, *SW* creates the correct reference by substituting the number of the keyed item or the page on which the marker appears in place of the key. For example, if you marked section 5.2.2 of your document with the key *part1*, you can create a reference to the section like this:

refer to section  ref: part1

The printed document will read "refer to section 5.2.2."

Note that *SW* substitutes only the number of the keyed item; you must remember to insert the appropriate identifier—such as "section," "page," or "chapter"— in your text before the reference.

## Citations

Citations are cross-references to bibliography items. You can cite reference materials by creating citations for items listed in your bibliography. In *SW*, you can create bibliographies manually or with BIBTEX. The processes for creating citations for the two kinds of bibliographies differ somewhat. However, both kinds of citations involve specifying keys for bibliography items and both kinds appear on the screen as a small gray box containing the word *cite* and the key you entered, like this:  cite: example . The citation is resolved when you print your document. Chapter 11 "Structuring Documents" contains information about creating both kinds of bibliographies and their citations.

## Jumping with Cross-references and Citations

Once you have created a cross-reference or a citation for a manually created bibliography item, you can use it to *jump,* or move the insertion point, to the beginning of the paragraph containing the specified marker.

---

**Note**   You can use a citation to jump to a manually created bibliography item but not to a BIBTEX bibliography item. See Chapter 11 "Structuring Documents" for more information about creating bibliography items and their keys.

---

▶ **To move to a marker specified in a cross-reference or citation**

- Hold down the CTRL key and click the mouse in the cross-reference or citation.
  –or–

1. Select the cross-reference or citation.

2. From the **Tools** menu, choose **Action** (ALT+L, A).

3. *SW* moves the insertion point to the paragraph containing the marker.

You can return to the cross-reference or citation. See "Retracing Your Steps" later in this chapter for more information.

▶ **To return to the cross-reference or citation**

- On the Navigate toolbar, click the History Back button ![icon].
  –or–
- From the Go menu, choose History Back (ALT+G, B).
  –or–
- Press CTRL+ALT+LEFT ARROW.
  The program moves the insertion point to the beginning of the paragraph containing the source of the previous jump.

# Using Hypertext

With *hypertext* you can jump immediately to a target elsewhere in your *SW* document. The target of a jump can be any numbered object with an identifying key, such as a figure, a section, a bibliography item, or an equation. *SW* uses hypertext to make *hypertext links,* which are jumps to specific targets in the same online document, and to make many other kinds of jumps:

- Jumps made from cross-references and citations.
- Jumps made with the commands on the Go menu (Previous Section, Next Section, Goto Marker, Goto Paragraph, or History Back)
- Jumps made with any of the buttons or the Section Heading area on the Navigate toolbar.

## Using Hypertext Links

Hypertext links are intended for documents that will be used primarily online. You can insert hypertext links throughout your online document to guide your readers to related topics, supplementary material, or key ideas. Hypertext links consist of two parts: a marker for the targeted item, which contains a key, and the *link* to the targeted item. On the screen, the hypertext link appears in green in the default screen font. In print, the link is replaced with the number of the targeted item, plus any text that you want to appear to its right and left.

For example, suppose you want to create a link to a section in your document that carries the number 3.2 and is called "Using Computational Software." You might create a hypertext link that appears like this in your online document:

For more information, see Using Computational Software.

When you print the document, the program will replace the link with the section number, so in print you will see

For more information, see 3.2.

When you create the link, you can specify text to appear on either side of the link, so that your printed text flows as smoothly as your online text. You might add the word *section* before the link and the words *later in this article* after the link. The appearance of the link on the screen is not affected, but its appearance in print will be different: when you preview or print the article, the program includes the text you specified and you will see this:

> For more information, see section 3.2 later in this article.

The program doesn't insert spaces around the link. If you want a space on the left or right of the link, you must specify it when you create the link. You can see examples of hypertext links in the *SW* online manuals and tutorial exercises.

## Creating Hypertext Links

▶ **To create a hypertext link**

1. On the Field toolbar, click the Hyper Link button 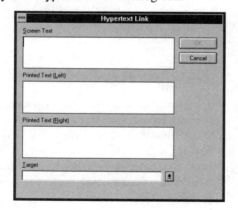.

   –or–

   From the Insert menu, choose Field and then choose Hypertext Link (ALT+I, E, H).

   The program displays the Hypertext Link dialog box.

2. In the Screen Text box, enter the text of the link as you want it to appear on the screen.

   When you print the document, this text will be replaced by the number of the targeted item.

3. In the Target box, enter the key of the targeted item.

   • If you want to create a link to an item that already has a marker, type the name of the marker.
     –or–

    a. Click the arrow next to the Target box to display the list of defined markers.

    b. Scroll the list to find the marker you want.

    c. Click the name to select it.

- If you want to create a link to an item that doesn't yet have a marker, enter the key for the item.

4. If you want to set the printed appearance of the link,

- In the Printed Text (Left) box, enter the text that you want to be printed before the link.
- In the Printed Text (Right) box, enter the text that you want to be printed after the link.
  Remember to include any necessary spaces on the left or the right of the link.

5. Choose OK.

## Jumping to a Hypertext Target

▶ **To jump to a hypertext target**

- Place the insertion point in the link and then from the Tools menu, choose Action (ALT+L, A).
  −or−
- Hold down the CTRL key and click the mouse in the hypertext link.

The program moves the insertion point to the beginning of the paragraph containing the specified marker.

## Retracing Your Steps

The History list is a record of the jumps you make within a document. *SW* activates the History list as soon as you make a jump of any kind, using hypertext links, the Go menu, the Navigate toolbar, cross-references, or citations. Each time you jump, the program adds an entry to the top of the History list. The entry shows the name of the section containing the paragraph to which you jumped. You can view the History list and use the list itself to jump to a target again.

▶ **To view the History list**

- From the Go menu, choose View History (ALT+G, H).
  The program displays the list.

### ▶ To jump to a section on the list

1. Open the History list.

2. Double-click the entry you want.

   The program moves the insertion point to the beginning of the section you chose.

### ▶ To return to the source of the most recent jump

- On the Navigate toolbar, click the History Back button .
  –or–

- From the Go menu, choose History Back (ALT+G, B).
  –or–

- Press CTRL+ALT+LEFT ARROW.
  The program moves the insertion point to the beginning of the paragraph containing the source of the previous jump.

## Using External Program Calls

An *external program* is a separate executable program. External program *calls* provide a way to start external programs from within an *SW* document. When you activate an external program call, you execute the targeted program. For example, without leaving your document, you can use external program calls to start demonstration programs, tutorials, or interactive examples.

External program calls appear on the screen in dark red in the default screen font. The call consists of a type, which includes a path that specifies the location of the program and specifies other details of how *SW* calls the external program. The call must also include text to specify how the call appears on the screen and, if the external program requires data, the name of the data file.

▶ **To insert a call to an external program**

1. Place the insertion point where you want the call.

2. On the Field toolbar, click the External Program Call button  .

   –or–

   From the Insert menu, choose Field and then choose External Program Call (ALT+I, E, E).

   The program opens the External Program Call dialog box.

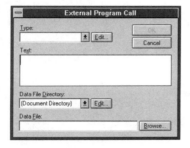

3. Click the arrow next to the Type box to display the available external program types.

4. Click the type you want.

5. In the Text box, enter the text of the call as you want it to appear in your document.

   The program displays this text on the screen in dark red letters in the default screen font.

6. If the external program requires a data file, enter the name and the directory of the data file.

7. Choose OK.

▶ **To add a type to the list of available external program calls**

1. On the Field toolbar, click the External Program Call button  .

   –or–

   From the Insert menu, choose Field and then choose External Program Call (ALT+I, E, E).

   The program opens the External Program Call dialog box.

2. Choose Edit (ALT+E) next to the Type box.

   The program opens the Call Type List dialog box. For each available type, the dialog box displays this information:

- **Path** - the location of the external program.
- **Parameters** - the information that must be sent to the external program.
- **Icon** - the location of the bitmap icon representing the external program.
- **Text Tag** - a tag that specifies the appearance of the call text on the screen.
- **Show As** - an indicator of whether the call should appear as an icon or text in the document.

3. Choose **Add** (ALT+A).

   The program opens the **Call Type** dialog box.

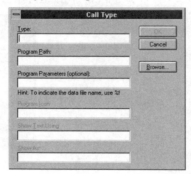

4. Enter the name and the path of the new type.

   The other information is optional.

5. Choose **OK** repeatedly to return to the document.

▶ **To execute an external program call**

1. Place the insertion point to the right of the call.

2. From the **Tools** menu, choose **Action** (ALT+L, A).

   –or–

   Hold down the CTRL key while you click the mouse in the call.

   The program executes the specified program.

# 7 Using Tables and Graphics

Use *SW* to create tables and import graphics that enhance the quality of your documents. You can specify the placement and appearance of tables and graphics, and you can add captions and keys so that you can refer to tables and graphics from elsewhere in your document.

## Creating Tables

A table consists of rows and columns of boxes called *cells* that you can fill with text, mathematics, and graphics. To make entering and editing table information easier, you can show the outline of the table cells on the screen by choosing **Matrix Lines** from the **View** menu. The outlines are intended as an editing guide; they do not print.

To position the insertion point in a cell, you click the mouse in the cell. To move from cell to cell, you use the TAB or ARROW keys. Otherwise, you move the insertion point within table cells in the same way you do elsewhere in a document.

You can edit tables to add or delete rows and columns, merge or split cells horizontally, or add lines to cells. You can also change the alignment and width of a column and the placement of the table in relation to the surrounding text. Finally, you can add captions and keys to certain kinds of tables so that you create cross-references to them.

### Entering Tables

Use the **Table** dialog box to specify the number of table rows and columns and to set the initial column and baseline alignment. You can set the baseline of the table, which aligns with the surrounding text, at the top or bottom row of the table or at its vertical center.

$$\text{Baseline on top row:}\quad \begin{matrix} 1 & 2 \\ 3 & 4 \end{matrix}$$

$$\text{Baseline at center:}\quad \begin{matrix} 1 & 2 \\ 3 & 4 \end{matrix}$$

$$\text{Baseline on bottom row:}\quad \begin{matrix} 1 & 2 \\ 3 & 4 \end{matrix}$$

▶ **To enter a table**

1. On the Standard toolbar, click the Table button  .

–or–

From the Insert menu, choose Table (ALT+F, A) and then press ENTER.

The program opens the Table dialog box.

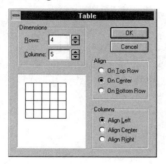

2. Specify the number of rows and columns you want in the table.

3. Specify the location of the table baseline.

4. Specify the column alignment for the table.

5. Choose OK.

## Changing the Table Dimensions

You can edit the dimensions of a table by adding, merging, or deleting cells; changing the width and alignment of cells; or adding lines to cells. You can delete the contents of a cell or cells, the contents of a row or column, and the row or column itself. Note, however, that if you select and delete the contents of an entire row or column in a single operation, that row or column is deleted from the table. To delete the contents of a row or column but keep the row or column in the table, delete the contents of at least one cell in a separate operation. The content of the cells is erased but the input boxes remain for you to refill.

▶ **To insert a row or column into a table**

1. Do one of the following:

   • Place the insertion point in the table or immediately to its right.
   • Select the entire table.
   • Select any group of cells.

2. From the Edit menu, choose Insert Row(s) (ALT+E, W) or Insert Column(s) (ALT+E, I).

The program displays the Insert Rows or Insert Columns dialog box.

3. Specify the number of rows or columns you want to insert.

If you selected a group of cells before you opened the dialog box, the number of rows or columns to add is set to the number of rows or columns you selected.

4. Specify the position of the insertion.

If you selected a group of cells before you opened the dialog box, the insertion arrow is shown to the left of or below the selection.

Note that the table illustration shown in the dialog box changes according to your specifications.

5. Choose OK.

► **To delete the contents of a cell**

1. Select the contents of the cell.

2. Press either DELETE or BACKSPACE.

   –or–

   From the Edit menu, choose Delete (ALT+E, D).

   The program removes the contents of the cell and replaces it with an empty input box.

► **To delete an entire row or column**

1. Select the entire row or column.

2. Press DELETE.

   –or–

   From the Edit menu, choose Delete (ALT+E, D).

► **To delete several columns**

1. Select at least one cell in each column you want to delete.

2. From the Edit menu, choose Delete Columns (ALT+E, E).

► **To merge cells horizontally**

1. Select the cells you want to merge.

2. From the Edit menu, choose Merge Cells (ALT+E, G).

Cells can be merged horizontally only. If your selection includes more than one row, cells in each row of your selection are merged.

▶ **To split previously merged cells**

1. Select the cells.

   - If you want to split a single cell that has previously been merged, select the contents of the cell.
   - If the cell is empty, place the insertion point in it.

2. From the Edit menu, choose Split Cells (ALT+E, S).

   *SW* splits the cells but leave all contents in the leftmost cells.

## Changing Table Properties

You can use the tab sheets in the Table Properties dialog box to change the alignment and width of columns, to change the baseline of the table, and to add lines to cells. Remember that when you choose OK or Cancel in a tabbed dialog box, you accept or discard the changes made on all tab sheets, not just the tab sheet in front.

▶ **To edit the alignment of a column**

1. Select the column or columns you want to align.

2. Open the Table Properties dialog box:

   - On the Standard toolbar, click the Properties button  .
     –or–
   - From the Edit menu, choose Properties (ALT+E, O).

3. Choose the Alignment tab to display the Alignment tab sheet.

4. In the Columns area, set the alignment you want.

5. Choose OK.

▶ **To edit the baseline alignment of the table**

1. Select the table or place the insertion point to its right.

2. Open the Table Properties dialog box.

3. Choose the Alignment tab to display the Alignment tab sheet.

4. In the Baseline area, select the alignment you want.

5. Choose OK.

▶ **To add or remove lines around cells**

1. Select the cells for which you want to add or remove lines.

   - If you want to modify the lines around a single cell, select the contents of that cell.
   - If the cell is empty, move the insertion point into it.

2. Open the Table Properties dialog box.

3. Choose the Lines tab sheet.

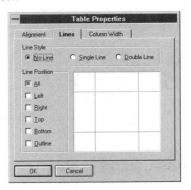

4. Select the line style.

5. Select the positions of the lines you want to add or remove.

   You can add or remove lines around any combination of the four sides of each cell in your selection.

6. Choose OK.

▶ **To set the column width**

1. Select the column or columns whose width you want to set.

   If your table has only one row, select the entire contents of the cell.

2. Open the Table Properties dialog box.

3. Choose the Column Width tab.

4. If you want the column to be wide enough to accommodate the widest entry in the column, check Use Automatic Width (ALT+A).

   The default setting for column width is Auto.

5. If you want to specify a column width

   a. Uncheck Use Automatic Width.
   b. Enter a value and a unit of measure for the width.

6. Choose OK.

   If you give a column fixed width, gray boxes appear in each of the cells of that column. To view or edit the contents of a cell, double-click the gray box. Text wraps within cells of a fixed-width column when you print.

## Positioning Tables

*SW* treats a table as if it were a single text character. A table can appear in a line with other text, like this,

| Head | Head | Head | Head |
|------|------|------|------|
| entry | entry | entry | entry |

, or in a centered paragraph of its own.

| Head | Head | Head | Head |
|------|------|------|------|
| entry | entry | entry | entry |
| entry | entry | entry | entry |

In addition, a table can *float*; that is, the table can be housed in a frame whose placement is determined when the document is printed.  A floating table can have captions, can be cross-referenced, and can be included in a list of figures or list of tables.

▶ **To display a table in its own centered paragraph**

1. Start a new paragraph.

2. Create a table.

3. Place the insertion point in the paragraph.

4. From the Section/Body Tag popup list, choose Body Center.

▶ **To create a floating table**

1. From the File menu, choose Import Fragment (ALT+F, R), then choose the table4_3 fragment from the Import Fragment dialog box and choose OK.

   –or–

   From the Fragments popup list, choose table4_3.

   *SW* inserts a table with four rows and three columns.  On the screen you see this:

|  | Head | Head | Head |  |
|---|---|---|---|---|
| [B] | entry | entry | entry | caption [E] |
|  | entry | entry | entry |  |
|  | entry | entry | entry |  |

2. Edit the dimensions of the table so that the table has the number of rows and columns you want.

3. Enter the contents of the cells.

## Labeling Tables

Floating tables can carry numbers, captions, and keys so that you can refer to them from elsewhere in your document.  If you add a caption, *SW* automatically creates the table number unless you suppress it.

▶ **To create a floating table with a caption and a key**

1. Import the fragment table4_3.

   *SW* inserts a table with four rows and three columns.  On the screen you see this:

| | Head | Head | Head | |
|---|---|---|---|---|
| [B] | entry | entry | entry | caption [E] |
| | entry | entry | entry | |
| | entry | entry | entry | |

2. Edit the dimensions of the table so that the table has the number of rows and columns you want.

3. Select the gray box marked *caption,* then choose **Properties**.

   –or–

   Double-click the gray box marked *caption*.

   The program opens the **TeX Field** dialog box, which shows this entry in the text area:

   \caption{Table Caption\label{key}}

4. If you do not want the table to be numbered, insert an asterisk after the word *caption*.

   Note that if the table doesn't have a number, you can't create cross-references to it.

5. Select the words *Table Caption* and then type the caption you want for the table.

6. If you want to add a key, select the word *key* and then type the key you want for the table.

7. Choose **OK**.

## Working with Graphics

You can import graphics from other applications into *SW*. Separate programs, called *graphics filters*, are used to import and display graphics created by other applications. Graphics in the following formats are supported:

| Format | Application | Format | Application |
|---|---|---|---|
| AI/EPS | Adobe Illustrator | JPG | JPEG |
| BMP | Windows bitmaps | PCT | Macintosh PICT files |
| CDR | Corel Draw files | PCX | ZSoft Paintbrush |
| CGM | Computer graphics metafiles | PIC | Lotus 1-2-3 |
| DRW | Micrografx Designer/Draw files | PS | PostScript Interpretive |
| DXF | AutoCAD drawing interchange | TIF | Tag image file format |
| EPS | Encapsulated PostScript files | TGA | Truevision TARGA |
| GIF | CompuServe Graphics | WMF | Windows metafiles |
| HGL | HP graphics language files | WPG | WordPerfect 2.0 |

The program automatically encloses each graphic file you import in a frame. By setting the properties of the frame in the **Graphic Properties** dialog box, you can change how the graphic is placed when you print your document. See the section "Placing Graphics in Frames" later in this chapter for more information. You can also use the tab sheets in the Graphic Properties dialog box to change the name and size of the graphic and to give the graphic a caption and a key. See Chapter 13 "Customizing SW" for more information about setting graphics defaults.

Remember that when you choose OK or Cancel in a tabbed dialog box, you accept or discard the changes made on all tab sheets, not just the tab sheet in front.

---

**Important**    Remember to use the *SW* Document Manager whenever you want to copy or send a document that contains graphics, so that the graphics files associated with the document are copied or sent along with the document. For more information, see Chapter 12 "Managing Documents."

---

## Importing Graphics

You can import an entire graphic or part of a graphic from another application.

▶ **To import an entire graphics file**

1. Move the insertion point to where you want the graphic.

2. From the File menu, choose Import Picture (ALT+F, I).

3. In the Import Picture dialog box, select the drive, directory, and file type of the graphics file you want to import.

4. In the File Name box, type or select the name of the file you want.

5. Choose OK.

   *SW* creates a link to the graphics file, which remains as a separate file in its original location. You should always use the *SW* Document Manager to move documents containing graphics so the graphics files are moved as well. See Chapter 12 "Managing Documents."

▶ **To paste a graphic or portion of a graphic from another application into an SW document**

1. Open the graphic from within the application you used to create it.

2. Select the portion of the graphic that you want to import.

3. Copy the selection to the clipboard.

4. Return to your *SW* document.

5. Position the insertion point where you want the graphic to appear.

6. From the Edit menu, choose Paste (ALT+E, P).

---

**Note**   If you paste graphics from the clipboard, *SW* stores your file as a Windows metafile and automatically generates a file name. The generated name does not describe the graphic. You should change the name of the file to something you understand as soon as possible.

---

▶ **To change the file name of a pasted graphic**

1. Select the graphic.

2. On the Standard toolbar, click the Properties button  .

   –or–

   From the Edit menu, choose **Properties** (ALT+E, O).

   The program opens the **Graphic Properties** dialog box.

3. Choose the **Picture Properties** tab.

4. Replace the name in the **File Name** box with a name you choose for the graphic and the extension `.wmf`.

5. Choose **OK**.

## Placing Graphics in Frames

A *frame* is a container that can hold a graphic or a plot. You determine the placement of the frame and its contents using the **Frame Properties** tab sheet in the **Graphic Properties** dialog box. The tab sheet has three placement options:

- *Inline placement* means the frame and its contents are just like a single text character, flowing with the text. The surrounding lines of text open up to accommodate the frame if it is higher or deeper than the text. You can enter an offset to set the baseline of the frame relative to the text baseline.
- *Displayed placement* is just like a mathematics display, centered in the line and ending the line above at the precise point you place the frame.
- *Floating placement* means the frame is displayed at a position determined at the time the document is printed. Floating frames can have captions, can be cross-referenced, and can be listed as part of a list of figures or list of tables. The floating placement options are as follows:
  - **Here** refers to the position in the text where the frame resides.
  - **Top** refers to the top of a page.
  - **Bottom** refers to the bottom of a page.
  - **On a Page of Floats** refers to a separate page containing no text, only other floating objects.

The Here option has precedence over the Top option. If all options are turned off, *SW* selects Top, Bottom, and On a Page of Floats as the defaults, in that order. You can select as many of the placement options as you want. Because the *SW* main window doesn't show printed pages, a frame with floating placement is shown on the screen as a display. When you preview or print the document, the frame and its contents are floated according to your placement request.

▶ **To set the frame properties**

1. Select the graphic or the frame.

   When the frame is selected, eight black handles appear around it.

2. On the Standard toolbar, click the Properties button .

   –or–

   From the **Edit** menu, choose **Properties** (ALT+E, O).

3. Choose the **Frame** tab to display the **Frame** tab sheet.

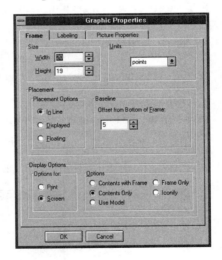

4. In the **Size** area, set the dimensions of the frame.

   When the tab sheet opens, it displays the current size of the graphic frame, in the units specified. The frame is slightly larger than the graphic itself

5. In the **Placement** area, choose the placement you want.

   The tab sheet changes depending on your choice.

   - If you choose **In Line**, you can set the amount you want the graphic to be offset from the baseline of the surrounding text.
   - If you choose **Floating**, you can choose the kind of placement you want.

6. In the Display Options area, choose the way you want the frame to appear in print and on the screen.

Note that if you choose Frame Only as a display option in print, you will be able to preview and print your document more quickly. When you are ready to produce a final printed version of your document, change the option to one that prints the graphic.

7. Choose OK.

## Resizing and Cropping Graphics

You can resize and crop a graphic with the mouse or more precisely using the Graphic Properties dialog box.

### Resizing Graphics with the Graphic Properties Dialog Box

▶ **To resize or crop a graphic using the Picture Properties tab sheet**

1. Select the graphic.

2. On the Standard toolbar, click the Properties button 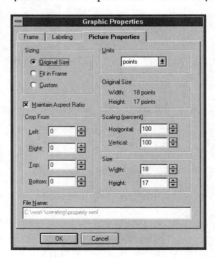.

   –or–

   From the Edit menu, choose Properties (ALT+E, O).

   The program opens the Graphic Properties dialog box.

3. Choose the Picture Properties tab to open the Picture Properties tab sheet.

4. In the Sizing area, select the basis on which the graphic is to be sized:

- Choose Original Size to use the graphic in the size it was imported.
- Choose Fit in Frame to scale the graphic to the frame size defined on the Frame tab sheet.
- Choose Custom to specify a size for the graphic.

The active unit of measure is shown at the top right of the tab sheet; click the Units arrow to select a new unit of measure.

5. Choose Maintain Aspect Ratio to keep the graphic in proportion as its size changes.

6. In the Crop From area, enter the amount you want to crop from each side of the graphic.

7. Set the size of the graphic:

- If you want to use a percentage of the original size, enter the horizontal and vertical percentages you want to use in the Scaling (percent) area.
- If you want to set the exact measurements, enter the width and height values in the Size area.

If you checked the Maintain Aspect Ratio box, both values in the set change proportionately when you enter the first one. If you didn't check the box, you must enter both values. Note too that the values in the Scaling and Size areas correspond. A change in one causes a change in the other.

8. Choose OK.


## Resizing Graphics with the Mouse

You can use the mouse to edit graphics by resizing them or by *panning* and zooming in and out to crop them. When you *pan* a graphic, you move a part of it outside the frame; only the part remaining in the frame appears on the screen or in print. When you *zoom in* on a graphic, the portion of the graphic that you select fills the entire frame. When you *zoom out*, the contents of the entire frame fits in the portion of the frame that you select.

Before you can use the mouse to edit a graphic, you must first select the graphic by clicking the mouse in the frame. When you select the graphic, eight small black handles appear around the frame, and a small Properties icon appears in the lower-right corner of the frame. If you click the box, the program opens the Graphic Properties dialog box. If you double-click in the frame, eight small gray handles appear around the frame and several other icons appear:

 When the mouse pointer is over the graphic, the pointer becomes a hand.

 Zoom In.

 Zoom Out.

 Graphic Properties.

Any sizing or cropping changes that you make with the mouse are reflected in the **Graphic Properties** dialog box.

### ▶ To resize a graphic using the mouse

1. Click the graphic.

2. Click and drag one of the handles around the graphic frame.

   As you drag the handle, the program displays a rubber band rectangle that indicates the new size. The resizing takes place when you release the mouse button.

### ▶ To pan a graphic using the mouse

1. Double-click the graphic.

2. Click and drag the graphic.

   As you drag, you move a rectangle with dotted sides. The rectangle represents the part of the graphic that you are moving outside the frame.

3. Release the mouse button.

### ▶ To zoom in

1. Double-click the graphic.

2. Click the Zoom In button .

3. Click and drag a rectangle in the frame.

   If **Maintain Aspect Ratio** is checked, the rectangle maintains its proportions as you drag it.

4. Release the mouse button.

   The part of the graphic you selected expands to fill the frame.

### ▶ To zoom out

1. Double-click the graphic.

2. Click the Zoom Out button .

3. Click and drag a rectangle in the frame.

   If **Maintain Aspect Ratio** is checked, the rectangle maintains its proportions as you drag it.

4. Release the mouse button.

   The part of the graphic that shows in the frame contracts to fill the rectangle you described.

## Labeling Graphics

Graphics can carry captions and keys. If you give a graphic a key, you can refer to it by number from elsewhere in the document.

▶ **To create a caption for a graphic**

1. Select the graphic.

2. On the Standard toolbar, click the Properties button 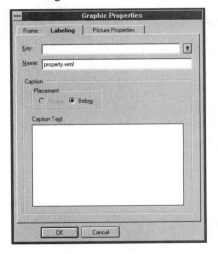.

    –or–

    From the **Edit** menu, choose **Properties** (ALT+E, O).

    –or–

    On the graphic, click the Graphic Properties button .

    The program opens the **Graphic Properties** dialog box.

3. Choose the **Labeling** tab.

    The program opens the **Labeling** tab sheet.

4. Check **Above** or **Below** to choose the placement of the caption.

5. In the **Caption Text** box, enter the caption for the graphic.

6. Choose **OK**.

▶ **To enter a key for a graphic**

1. Select the graphic.

2. On the Standard toolbar, click the Properties button  .

   –or–

   From the Edit menu, choose Properties (ALT+E, O).

   –or–

   On the graphic, click the Graphic Properties button  .

   The program opens the Graphic Properties dialog box.

3. Choose the Labeling tab.

   The program opens the Labeling tab sheet.

4. In the Key box, enter the key for the graphic.

   If you want to see a list of keys already defined for the document, click the arrow next to the Key box.

5. Choose OK.

# 8 Using the Find and Replace Feature

With the Find and Replace commands on the Edit menu, you can quickly locate or change text or mathematics in your document. In *SW*, you can

- Find all occurrences of a specified word, phrase, sequence of characters, mathematical expression, or combination of mathematics and text.
- Find text or mathematics with a specific tag, such as Emphasize or Section head.
- Replace each occurrence of the specified text or mathematics with different text or mathematics.
- Replace the tags applied to each occurrence of the specified text or mathematics.

We suggest you save your document before you use Replace. If you don't like the results, you can close the document without saving the changes.

## Specifying Search and Replacement Patterns

When you choose the Find or Replace command, *SW* opens a dialog box in which you can specify the search and replacement patterns and the options that govern how the program searches your document. You enter text and mathematics into the search and replacement patterns exactly as you enter them into a document in the main window.

The program retains the search and replacement patterns until you change them, unless you specify that they should not be retained from one work session to the next. To specify something different from what appears in the Search for or Replace with box, first delete the old patterns and then enter the new ones.

▶ **To discard search and replacement patterns when you close a session**

1. From the Tools menu, choose User Setup (ALT+L, U).

2. Open the General tab sheet.

3. In the General Options area, uncheck the Save Find/Replace Text check box.

4. Choose OK.

## Searching for Text

The menu commands and toolbars are active when the Find and Replace dialog boxes are open, so you can type text just as you do when you enter text in a document. You can also use the Copy and Paste commands to copy text from the main document to the dialog boxes.

If your search pattern is in text, the program searches both the text and mathematics for occurrences of the pattern. Ordinarily, a text search locates the characters you specify anywhere they occur, whether they stand alone or appear as part of a word. For example, if you search for the text characters *on*, *SW* will find *on*, *tone*, and *Ontario*. You can use two search options—Whole Words Only and Exact Case—to define a text search more precisely.

- The Whole Words Only option locates specified text characters only when they appear as a separate word.
  For example, if you want to locate the text characters *on* and specify Whole Words Only, the program will locate only the word *on*.

- The Exact Case option locates specified text characters only when they match a particular pattern of capitalization.
  For example, by entering *Brown* and specifying Exact Case, the program will locate the name *Brown* but bypass uncapitalized references to the color *brown*.

## Searching for Mathematics

The toolbars and menu commands are active when the Find and Replace dialog boxes are open, so you can type mathematics just as you do when you enter mathematics in a document. You can also use the Copy and Paste commands to copy mathematics from the main document to the Find and Replace dialog boxes.

If the search pattern is in mathematics, the search is limited to the mathematics in the search area. That is, if your search pattern is an $x$ in mathematics, the program searches only the mathematics in the search area, bypassing words such as *exit* and *hexadecimal*.

If you specify an empty mathematical template as the search pattern, the program finds all occurrences of that template in the search area. For example, if you enter an empty fraction, the program finds all fractions in the search area. You can enter an empty matrix of any shape to search for all matrices, regardless of their shape.

*SW* uses a search method that finds matches on the initial part of a word. Because *SW* treats mathematical objects as words, you may find that a search for a mathematical expression yields unexpected results. For example, the expression $x^2$ consists of two words, the variable and the superscript. If you specify $x^2$ as the search pattern, the program will find $x^{2y}$, $x_n^2$, and $x^2$. Use the Whole Words Only option to locate an exact match of a mathematics search pattern.

If you specify a replacement pattern for all occurrences of a mathematics search pattern and the mathematical objects found in the search do not match the search pattern exactly, *SW* asks you to confirm each replacement.

## Using Tags in Search and Replacement Patterns

The tag popup lists are active when the Find and Replace dialog boxes are open, so you can use the tag popup lists to apply tags to the search and replacement patterns. For example, you can replace every occurrence of the emphasized phrase

*for all real $\Gamma$*

with the phrase

for every real $\Lambda$

or you can change all occurrences of $C$ to $\mathcal{C}$. If you apply a tag to the search pattern, the program locates only those occurrences that carry the same tag. For example, if you specify the word *step* with the subheading tag, the program locates the word *step* every time it appears in a subsection heading but bypasses the word anywhere else in the document. If your search pattern is not tagged, the program will find all occurrences of the pattern in the search area, whether tagged or untagged.

If you apply a tag to an empty search pattern—that is, choose a tag but enter no text in the search pattern—the program will find all occurrences of the tag in the search area. For example, you can search for all paragraphs in the document that are tagged as section heads. Similarly, you can replace all occurrences of a certain tag with another tag of the same type, such as replacing one kind of text tag with another. For example, you can replace all bold text tags with emphasize text tags, all subsubsection heads with subsection heads, or all bulleted list items with numbered list items.

# Specifying the Search Area

Unless you first specify a search area by selecting a part of your document, the program searches and replaces throughout the main part of the document. If you want to search a part of your document that appears in a dialog box—such as a footnote—you must open the dialog box before you initiate the search. Searching begins at the insertion point.

You can search your document in both directions. Ordinarily, the program searches forward from the insertion point to the end of the document. If you specify the Search Backward option, the program searches backward, locating only those occurrences of the search pattern that appear between the insertion point and the beginning of the document. The Search Backward option is not available when you search a selection.

If you begin the search from the middle of your document, the program searches from the insertion point to the end or, if you choose Search Backward, from the insertion point to the beginning. You are then asked whether you want to continue the search in the rest of the file or stop.

## Finding and Replacing

▶ **To find text or mathematics**

1. If you want to search part of the document rather than all of it, select the part you want to search.

2. From the Edit menu, choose Find (ALT+E, F) to open the Find dialog box.

3. Enter the text or mathematics you want to find in the Search for dialog box.

4. Select any options you need to control the search.

5. From the Find menu, choose Find Next (ALT+N) to begin searching.

   The program selects the first occurrence of the search text or mathematics and shows it at the top of the screen.

6. To find the next occurrence, choose Find Next (ALT+N) again.

   The program searches from the insertion point to the end (or to the beginning, if you choose Search Backward) and asks if you want to continue the search.

   • Choose Yes to search through the remainder of the document.
     –or–
   • Choose No to stop the search.

▶ **To make changes to a found occurrence**

1. Choose Cancel.

2. Edit the text or mathematics.

3. To resume the search, choose Find from the Edit menu again (ALT+E, F) and then choose Find Next (ALT+N).

▶ **To replace text or mathematics**

1. Save your document.

   If you don't like the results of the Replace operation, you can close the document without saving changes.

2. If you want to search and replace in part of the document rather than all of it, select the part you want to search and replace.

3. From the Edit menu, choose Replace (ALT, E, R).

The **Replace** dialog box appears. If the **Find** or **Replace** command has been used in the current session, the last search pattern specified appears in the **Search for** box. If the **Replace** command has been used in the current session, the last replacement text specified appears in the **Replace with** box.

4. Enter the text or mathematics you want to find in the **Search for** box.

5. Enter the replacement text or mathematics in the **Replace with** box.

6. Select any options you need to control the way changes are made in your document.

7. To begin searching:

- Choose **Replace All** (ALT+A) if you want to replace all occurrences of the search text with the replacement text without confirming each replacement.
- Choose **Find Next** (ALT+N) if you want to confirm the replacement of each occurrence of the search text.
  When the program finds an occurrence of the search text

  - Choose **Replace** (ALT+P) to make the change.
    −or−
  - Choose **Find Next** (ALT+N) to leave the occurrence unchanged and continue the search.

# 9 Checking Spelling

In *SW,* you can check the spelling of all or part of your document using a dictionary supplied with the program and a supplementary user dictionary whose contents you determine.

We ship each system with one language dictionary. You can purchase additional dictionaries directly from TCI for many languages, including American English, British English, Danish, Dutch, Finnish, French, French Canadian, German, Swiss German, Italian, Norwegian (Bokmal), Norwegian (Nynorsk), Brazilian Portuguese, Continental Portuguese, Spanish, Catalan, and Swedish.

You can check the spelling of a single word or the spelling in a selection, in an *SW* text box within a dialog box, in the portion of the document between the insertion point and the end of the document, or in the entire document. Unless you have made a selection, you can start the spell check at the beginning of your document, regardless of the position of the insertion point. The program does not check spelling in mathematics, nor does it check words embedded within mathematics.

If you check 20 or fewer words, the program displays the number of words checked in the status area at the bottom of the screen. If you check more than 20 words, a dialog box pops up giving you the number of words checked and the number of replacements made.

---

**Note**   You can use the spell check feature to count the number of words in your document by starting the spell check at the beginning of the document. This count does not include mathematics or words in mathematics.

---

You establish the kind of spell check you want using the **Spell Check** dialog box. The choice to check the spelling from the beginning of the document must be set each time you open *SW.* The program retains the other choices you make and uses them as defaults until you change them.

▶ **To set the Spell Check options**

1. Make sure that no selection has been made and you can see the blinking insertion point.

2. On the Standard toolbar, click the Spelling button  .

   –or–

   From the **Tools** menu, choose **Spelling** (ALT+L, S).

   The program opens the **Spell Check** dialog box.

3. Use the controls in the dialog box to specify the options you want:

- Check the **Beginning of Document** box to begin the current spell check at the beginning of the document.
- Choose the **Language Options** button to set the language to be used in the spell check for all documents or for the current document only. (This option is relevant only if you have more than one language installed.)
- Choose the **Edit Dictionary** button to modify the user dictionary by adding or removing words.
- Choose the **Options** button to establish checks for repeated words, words containing numbers, and words with initial capitals.

We suggest you save your document before you check the spelling in your document. If you don't like the results, you can close the document without saving the changes.

## Checking the Spelling in a Document

▶ **To check the spelling of the entire document**

1. On the Standard toolbar, click the Spelling button .

    –or–

    From the **Tools** menu, choose **Spelling** (ALT+L, S).

2. In the **Spell Check** dialog box, check **Beginning of Document**.

3. If you want to change the spell check options, choose **Options** (ALT+O).

    The program displays the **Spell Check Options** dialog box.

4. Check the boxes for the options you want.

5. Choose **OK** to close the **Spell Check Options** dialog box.

6. Choose **OK** to begin the spell check.

▶ **To check the spelling of the document from the insertion point to the end of the document**

1. Place the insertion point in the body of your document where you want the spell check to begin.

2. On the Standard toolbar, click the Spelling button .

   –or–

   From the Tools menu, choose Spelling (ALT+L, S).

3. Make sure the **Beginning of Document** check box is not checked.

4. Change the options as necessary in the **Spell Check** dialog box.

5. Choose **OK**.

▶ **To check the spelling in a selection**

1. Select the part of the document you want to check.

2. On the Standard toolbar, click the Spelling button .

   –or–

   From the Tools menu, choose Spelling (ALT+L, S).

   The spell check begins immediately.

   Each time *SW* finds a word that is not in the dictionary, it opens a dialog box that indicates the word and lists spelling alternatives. You have the option to choose a correction and replace the word once or each time it is encountered, skip the word once or every time it is encountered, add the word to the user dictionary, or cancel the spell check.

▶ **To correct a misspelled word**

- If the **Alternatives** list contains the correct form of the word
  - Double-click that word to replace the current occurrence.
    –or–
  - Select the word and choose **Replace All** to replace all occurrences.
- If the **Alternatives** list doesn't contain the correct form of the word
    a. Enter the correct spelling in the **Replace with** box.
    b. Choose **Replace** to change only the current occurrence of the misspelling.
       –or–
       Choose **Replace All** to replace all occurrences of the misspelling.

  If you choose **Replace All**, it is valid only to the end of the current spell check operation.

▶ **To skip a questioned word**

- Choose **Skip** to leave this occurrence as it is.
  –or–
- Choose **Skip All** to skip all further occurrences of the word.
  If you choose **Skip All**, it is valid only to the end of the current spell check operation.

▶ **To add a questioned word to the user dictionary**

- Choose **Add to Dictionary**.

  The word will remain in the user dictionary until you remove it. See "Modifying the User Dictionary" later in this chapter.

▶ **To remove a repeated word from the document**

1. Press DELETE to delete the word from the dialog box.

2. Choose **Replace**.

▶ **To stop the spelling check**

- Choose **Cancel**.

## Modifying the User Dictionary

You can modify the contents of your user dictionary by adding or deleting words. This feature is useful for including unusual spellings of certain words.

---

**Tip**   You may need to remove the unusual spellings from the User Dictionary when you're ready to work in another document.

---

▶ **To modify your user dictionary**

1. Open the **Spell Check** dialog box.

2. Choose **Edit Dictionary** (ALT+D).

   The system opens the **User Dictionary** dialog box.

3. If you want to add a word

    a. Type the word in the top box.
    b. Choose **Add** (ALT+A).

4. If you want to remove a word

    a. Select the word from the list in the bottom box.
    b. Choose **Remove** (ALT+R).

5. When you have finished modifying the dictionary, choose **OK**.

6. Choose **OK** again to start a spell check.

    –or–

    Choose **Close** to return to the main window.

▶ **To modify the user dictionary during a spell check**

1. Start a spell check.

2. When a word that is not in the internal dictionary is located, choose **Add to Dictionary** from the **Spell Check** dialog box.

The system accepts the spelling of the word and adds it to the user dictionary. The program continues the spell check, accepting all other occurrences of the word you added.

## Installing a Language Dictionary

*SW* must be installed before you can add additional language dictionaries. The installation program for additional dictionaries is a Windows program.

---

**Important**    Please read the license agreement that comes with each *SW* language dictionary. By opening the disk package, you accept the terms of the agreement.

---

▶ **To install an additional language dictionary on your hard drive**

• Follow the instructions on the language dictionary disk.

## Specifying a Spell Check Language

If you have dictionaries for more than one language installed on your system, you can specify the dictionary you want to use to check the spelling of all new documents and the current document.

► **To specify a spell check language dictionary**

1. Open the Spell Check dialog box.

2. Choose Language Options.

    The system opens the Spell Check Language Options dialog box.

3. In the Default Language for New Documents box, select the language dictionary you want to use as the default for checking the spelling of all new documents.

4. In the Language for Current Document box, select the language dictionary you want to use for checking the spelling of the current document.

5. Choose OK.

# 10  Applying Styles to Documents

In *SW*, the printed appearance of your document differs from the screen appearance, although both are determined by the *typesetting style* you choose. Typesetting styles are collections of commands that define page size, typeface, font size, margins, indentation, page numbering, and many other details of typography and layout. Many typesetting styles define the presence of certain structural elements, such as tables of contents, indexes, and heading levels. Styles also determine how various document elements, such as section headings and body text, are displayed on the screen.

*SW* styles are designed to produce beautiful printed documents characterized by precise and consistent formatting of text and mathematics. Line breaking is sophisticated, with words automatically hyphenated when line breaks demand it. The fonts used for printing have been selected for readability and have kerning and ligatures added where necessary.

Typesetting styles are also designed to make reading, editing, and interacting with your on-screen document as easy as possible. Styles break lines to fit on the screen, regardless of the size of the main window, so you rarely need to scroll horizontally. The styles use screen fonts that have been selected for ease of readability and are normally different from those used in print.

*SW* is distributed with a large selection of styles for articles, reports, books, letters, and other *document types*. Each style has a *base style*, a file of instructions that define the basic style elements of the corresponding document type. Each style also has a series of *formatting options*. When you choose a style, the format of the document is determined automatically. You can, however, tailor the style to your needs by changing such formatting options as the body text font size, paper size, page orientation, and number of text columns. Some styles also use *packages* that enable *SW* to customize your document in some way, such as by choosing the language to be used for typesetting, including certain kinds of fonts or graphics, or creating an index. Packages can have options also.

Not all typesetting styles are complex. When all you need is a simple document, choose one of the Easy Styles provided with the program. Easy Styles are simple, predefined styles for common types of documents including letters, memos, faxes, exams, reports, articles, and books. All Easy Styles are identified with the word *Easy* in the name. These styles are intended to simplify the process of writing a quick document when complex formatting is not an issue. Each Easy Style contains a complete *document shell* consisting of fields containing predefined information. To work with a shell, all you do is replace the shell's information with your own. Chapter 11 "Structuring Documents" contains more information on using document shells.

If you need to produce documents using $\mathcal{AMS}$-LaTeX styles, follow the instructions in "Choosing a Style" later in this chapter to choose either the AMS Article Style from the math article styles or AMS Book Style from the book styles. These styles are listed in the **Predefined Styles** dialog box when LaTeX $2_\varepsilon$ is selected.

To find the typesetting style that best meets your requirements, you can preview sample documents typeset with the styles provided with *SW*. When you're ready to produce final printed output, you can check the typeset appearance of your document with the previewer. Before you print you may want to make minor adjustments such as inserting specific page breaks, additional vertical spacing, or horizontal lines. We recommend you leave these adjustments for the last stage of document preparation so that you do the work only once.

If you find you need to make major adjustments to the printed appearance of your document, you may need to tailor the style options further or choose a style that more closely matches your requirements. If you can't find predefined styles appropriate for your documents, you can use the *SW* Style Editor to define your own styles. The online Help for the *SW* Style Editor explains in detail how to create styles. If you are familiar with LaTeX, you can "Go Native" to use a LaTeX style that is not predefined in *SW*.

## Choosing a Typesetting Style

You can choose a style, change the style's formatting options and packages, or change to a different style at any point in the creation of your document. We recommend, however, that you choose the style carefully before you create the content of your document. Preview the available styles, and choose the one that best meets your needs.

---

**Important**   Changing from one style to another is most successful if both styles produce the same type of document and use the same base style. Changing from one type of document, such as a book, to a very different type of document, such as a letter, can produce unpredictable results. See "Changing the Style" later in this chapter for more information.

---

## Previewing a Style

Your *SW* software includes a set of sample documents that have been typeset with different styles. You can preview these sample documents to find the typesetting style that best meets your requirements.

▶ **To preview a style**

1. From the File menu, choose Style (ALT+F, Y).

   The program displays the LaTeX Styles for Print and Preview dialog box, which names the currently selected style.

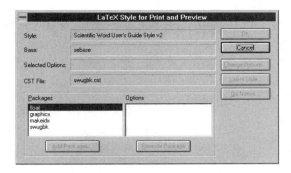

2. Choose **Select Style** (ALT+S).

   The program displays the **Predefined Styles** dialog box:

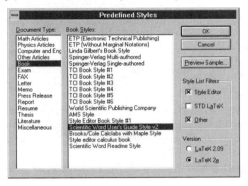

3. Select the document type and style you want to preview.

4. If you want to list more styles, check additional options in the **Style List Filters** area.

5. Choose **Preview Sample** (ALT+P).

   The program opens the previewer to display a sample document that has been typeset using the style you selected. Scroll through the document to examine the typeset appearance of the style.

6. From the **File** menu in the previewer, choose **Exit** (ALT+F, X).

7. In the **Predefined Styles** dialog box

   - Choose **OK** to accept the previewed style and return to the **LaTeX Styles for Print and Preview** dialog box.
     –or–
   - Choose **Cancel** to leave the current style unchanged.
     –or–
   - Repeat steps 3–6 to preview another style.

8. From the **LaTeX Styles for Print and Preview** dialog box

- Choose **OK** to accept the style shown in the dialog box.

  –or–

- Choose **Cancel** to make no changes to the current style.

## Choosing a Style

*SW* is distributed with a large selection of styles for articles, reports, books, letters, and other document types. Choose the style that is best for your document.

▶ **To choose a style**

1. From the **File** menu, choose **Style** (ALT+F, T).

   The program displays the **LaTeX Style for Print and Preview** dialog box.

2. Choose **Select Style**.

   The program displays the **Predefined Styles** dialog box. The panel on the left of the dialog box lists the document types for which typesetting styles are available. The panel on the right lists the styles for the document type currently selected.

3. Select the type of document you want from the panel on the left.

4. Select the style you want from the panel on the right.

5. If you want to list more styles, check additional options in the **Style List Filters** area.

6. In the **Version** area, choose the version of LaTeX you want *SW* to use in formatting your document.

   You can choose LaTeX $2_\varepsilon$, the newest version of LaTeX, or LaTeX2.09, an earlier version. Choose LaTeX $2_\varepsilon$ unless you have a specific need for the earlier version. For example, you might need to send a document to a colleague who does not have the latest release of *SW* or has only the earlier version of LaTeX.

7. Choose **OK** to accept the document type and style.

   The dialog box now shows the currently selected style and base, the name of the corresponding `.cst` file, and any formatting options or packages in effect. The `.cst` file controls a wide range of screen behavior and tag choices and also contains information that the LaTeX filter uses when reading and writing a `.tex` file.

8. Choose **OK** to leave the **LaTeX Style for Print or Preview** dialog box.

## Changing the Style Options

Although the style you choose determines the format of your document, you can tailor some aspects of the style from within *SW*. By changing the formatting options, you can alter such things as the font size for body text, the size of the paper, the orientation of the page, and the numbering style for theorems. By changing the options for the packages associated with the style, you can further customize the appearance of your document.

▶ **To change the style options**

1. From the File menu, choose Style (ALT+F, Y).

   The program displays the LaTeX Style for Print and Preview dialog box. The dialog box shows the selected style and any options currently in effect.

2. Choose Change Options (ALT+C).

   The program opens the Style Options dialog box.

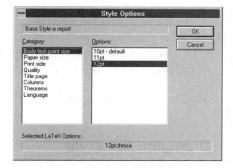

The left-hand panel lists the formatting elements for which you can specify options from within *SW*. The elements differ for each style. The right-hand panel lists the available options for the selected element. The box on the bottom displays all selected options.

3. Choose a formatting element for which you want to specify an option.

4. Select the option you want.

5. Repeat steps 3 and 4 until you have set all the options you want.

6. Choose OK to leave the Style Options dialog box.

7. Choose OK to leave the LaTeX Style for Print or Preview dialog box.

▶ **To change the packages**

1. From the File menu, choose Style (ALT+F, Y).

   The program displays the LaTeX Style for Print and Preview dialog box. The dialog box shows the selected style and any packages currently in effect.

2. If you want to add a package to the style, choose Add a Package (ALT+A).

   The program displays the Package Options dialog box.

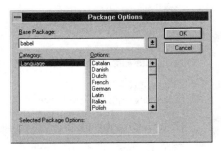

a. Click the arrow to the right of the **Base Package** box for a list of the packages available for the style.

b. Select the package you want.

c. If options are listed for the package, select the option you want.
   Your choices are reflected in the box at the bottom.

d. Choose **OK** to leave the **Package Options** dialog box.

3. Choose **OK** to leave the **LaTeX Style for Print or Preview** dialog box.

## Changing the Style

You can change the printed appearance of your document by changing the typesetting style you assign to it. Changing from one style to another is most successful when both styles produce the same type of document, such as a book or a report. Styles in the same document type generally contain the same type of structural elements. For example, books generally contain a table of contents, an index, and a bibliography and may also contain a list of figures or tables, a preface, a page of acknowledgments, and several appendixes. Letters, on the other hand, contain none of those elements but do contain a salutation and a closing, and depending on their formality, an inside address, a subject line, and enclosure indicators.

Because the structural elements inherent in one document type may not transfer smoothly to another document type, changing the style of a document to a style of a different type can produce unpredictable results and may even damage your document.

If you want to change the style of a document to a different type, you should create a new document file in the style you want and then import the contents of the original document into the new file, as described in the instructions later in this section.

---

**Important**   Changing from one style to another in *SW* is most successful if both styles produce the same type of document. Changing from one type of document to another, such as from an article to a memorandum, can damage your document.

---

▶ **To change to a style with the same document type**

1. From the **File** menu, choose **Style** (ALT+F, Y).

2. Choose **Select Style**.

3. From the **Styles** box, choose the style you want.

4. Choose **OK** to leave the **Predefined Styles** dialog box.

5. Choose **OK** to leave the **LaTeX Style for Print and Preview** dialog box.

Depending on the styles in question, the program displays a message about saving and reloading your document.

6. Choose **OK**.

*SW* saves and reloads your document, which looks the same in the main window as it did before. Because both the old and the new styles produce the same type of document, you do not have to give the program any further formatting instructions.

▶ **To change to a style with a different document type**

1. Save a copy of your document.

    a. From the **File** menu, choose **Save As** (ALT+F, A).
    b. In the **Save As** dialog box, enter a name for the copy of your document.
      You can assign any acceptable DOS file name to the copy of your document.
    c. Choose **OK**.

2. Create a new document file with a different style.

    a. On the Standard toolbar, click the New button .
      –or–
      From the **File** menu, choose **New** (ALT+F, N).
    b. From the **Predefined Styles** dialog box, choose the document type you want, then choose the style.
    c. Choose **OK**.

3. Import your original document into the new document file.

    a. From the **File** menu, choose **Import Contents** (ALT+F, T).
    b. In the **Import Contents** dialog box, enter the name and directory of the copy of your document.
    c. Choose **OK**.
      *SW* imports the body of your document into the new document file and asks if you want to import the front matter of the original document.
    d. Choose **Yes** to import the front matter and add it to any front matter already existing in the active document.
      –or–
      Choose **No** to leave the front matter of the active document unchanged.

4. Preview the document to see how your document will appear in print.

## Going Native

If you have a LaTeX style that you want to apply to your document, you can force *SW* to use it to typeset your document. In this case, the program bypasses all predefined styles and, when you preview or print your document, passes your style input directly to LaTeX for processing.

---

**Important**   You should be very familiar with LaTeX styles and options if you choose to use a style of your own.

---

▶ **To use a LaTeX style not predefined in SW**

1. Copy the style to the `tex\macros` directory in the main *SW* directory (normally `swp25`).

2. From the File menu, choose Style (ALT+F, Y).

   The program displays the LaTeX Style for Print and Preview dialog box. The dialog box shows the currently selected style and options.

3. Choose Go Native.

   The program opens the Native LaTeX Style dialog box.

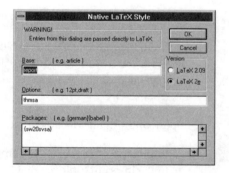

4. In the Base box, enter the document type.

5. In the Options box, enter any options you want to define for the style.

6. In the Packages box, enter the packages appropriate for the style.

   The options and packages you specify will be input directly into LaTeX.

7. Choose OK to leave the Native LaTeX Style dialog box.

8. Choose OK to leave the LaTeX Style for Print or Preview dialog box.

# Creating a Typesetting Style

In *SW*, you can create typesetting styles in two ways.

- Use the *SW* Style Editor to modify an existing Style Editor style.
  Only styles created with the Style Editor styles can be modified with the Style Editor. You can list the styles that are Style Editor styles by opening the **Predefined Styles** dialog box, unchecking **STD LaTeX** and **Other** in the **Style List Filters** area, and then checking **Style Editor**.
  The best way to create a new style is to open an existing style with the Style Editor, save it with a new name, and then make your modifications. Styles you create with the Style Editor are automatically added to the list of predefined styles. See the online Help available from the *SW* Style Editor for more information.

- Use LaTeX to create new typesetting styles.
  If you're very familiar with LaTeX and in particular with styles, you can create styles in LaTeX and add them to the list of predefined styles. The Technical Reference, supplied as `techref.tex` in the `extras` directory, contains more information.

# Making Final Adjustments to the Printed Appearance

The automatic features of the predefined typesetting styles ensure that many documents need no final adjustments to line breaks. Occasionally, however, you may want to specify changes to the appearance of certain paragraphs or pages. In *SW*, you can specify breaks in lines, pages, and mathematical expressions, and you can add additional vertical or horizontal space or insert horizontal lines, or *rules*. Larger articles and books inevitably need some final adjustments. This section describes how to make those adjustments.

---

**Note**   If you're new to *SW*, we encourage you focus on creating and refining the content of your document instead of its appearance. Experiment with various styles to discover those that best suit your needs. If you find that occasional formatting adjustments are still necessary, follow the instructions in this section.

---

▶ **To prepare a large document for final printing**

1. Complete all work on the content of the document.

   You may find it difficult to do this if you're new to *SW*, but developing the habit is worthwhile. Your productivity will increase markedly as a result.

2. Preview the document and check carefully for any cramped horizontal or vertical spacing, mathematics too large to break properly, or bad line or page breaks.

3. Starting from each forced page break in your document (for example, at the beginning of each chapter), make spacing adjustments according to the instructions that follow.

4. After each series of adjustments, save and preview the document.

5. Print the document.

## Adjusting Horizontal Space

Although all $SW$ styles adhere to international mathematics typesetting standards, you may occasionally want to add additional horizontal space between words or elements on a line. You can alter the amount of space between text or mathematical characters using the Spacing command on the Insert menu. These spacing options are available:

- Em Space adds a space the width of the letter M.
- 2-Em Space adds a space the width of the letters MM.
- Normal Space adds a space of $\frac{1}{6}$ em.
- Required Space adds a space of $\frac{1}{6}$ em.
- Non-breaking Space adds a space of $\frac{1}{6}$ em.
- Thin Space adds a space of $\frac{2}{9}$ em.
- Thick Space adds a space of $\frac{5}{18}$ em.
- Italic Correction adds a space whose width depends on the character to its left.
- Negative Thin Space adds a space of $-\frac{2}{9}$ em.
- Zero Space adds no space.
  Use a zero space when you create prescripts, as described in Chapter 4 "Entering and Editing Mathematics."
- Custom sets the additional horizontal spacing to any value you want.
  You can fill certain kinds of custom space with nothing, with dots, or with a line. You can discard the space if it falls at the end of a line, or require that the space be included regardless of where it falls on the page.
  See Chapter 3 "Entering and Editing Text" and Chapter 4 "Entering and Editing Mathematics" for more information.

---

**Note**    LaTeX automatically inserts the proper spacing when it typesets your text. However, sometimes you may not want space where LaTeX wants to insert it. For example, the space that follows an abbreviation in the middle of a sentence should be less than the space that follows a period at the end of a sentence. You can prevent end-of-sentence spacing by inserting a required space or a nonbreaking space.

---

▶ **To insert horizontal space on a line**

1. Position the insertion point where you want the space to begin.

2. On the Math toolbar, click the Space button [e.m] .

   –or–

   From the Insert menu, choose Spacing, and then choose Horizontal Space (ALT+I, S, H).

3. Choose the space you want.

4. Choose OK.

▶ **To insert custom horizontal space**

1. Position the insertion point where you want the space to begin.

2. On the Math toolbar, click the Space button  .

   –or–

   From the Insert menu, choose Spacing, and then choose Horizontal Space (ALT+I, S, H).

3. Choose Custom.

4. Specify whether you want the space to be *fixed* or *stretchy*

   - Choose Fixed if you want LaTeX to insert an exact amount of space.

     a. In the Width box, specify the amount of space you want inserted.
     b. In the Units box, specify the unit of measure.

   - Choose Stretchy if you want LaTeX to determine the amount of space that is inserted when the document is typeset.
     The amount is based on a *stretch factor,* which you specify. A stretch factor of 1 fills the current line.

     a. In the Factor box, specify the stretch factor.
     b. If you specify a factor of 1, you can specify that the space should be filled with a line, dots, or nothing.

5. Specify whether you want the space to be inserted always or whether it can be discarded if it falls at the end of a line.

6. Choose OK.

   The program inserts the space you specify. If you have the Invisibles option turned on, the space appears on the screen as a horizontal line, which does not appear in print.

## Adjusting Breaks in Mathematical Expressions

If an expression extends into the margin because it is too large to break properly, we suggest you display it. If you don't want to display the expression, you can indicate places where a break is appropriate, required, or not allowed.

In general, it is preferable to suggest a break rather than to force one. If a break is suggested rather than forced, it may not occur if later changes move the expression to a position where a break is no longer necessary.

▶ **To adjust a break**

1. Move the insertion point to where you will allow a break.

2. From the Insert menu, choose Spacing (ALT+I, S).

3. Choose Break (ALT+B).

4. Choose the break option you want:

   - If you want to suggest a break, choose Allowbreak (ALT+A).
   - If you want to force a break, choose Newline (ALT+N).
   - If you want to prevent a break, choose No break (ALT+N).

5. Choose OK.

## Adjusting Vertical Space

The style determines the line spacing.  Some styles provide a double-spacing option. Occasionally, you may want to create additional vertical space between paragraphs or between lines and other elements. You can add vertical space with the Spacing command on the Insert menu and with several fragments:

- Small Skip adds a space of about $\frac{5}{16}$ inch between lines in a 12-point style.
- Medium Skip adds a space of about $\frac{3}{8}$ inch between lines in a 12-point style.
- Big Skip adds a space of about $\frac{7}{16}$ inch between lines in a 12-point style.
- Strut forces the maximum line space required by the font.
  The space increases the consistency of line spacing in situations where the spacing is dependent on the letters in the text, such as in cells in a table.

- Math Strut adds a space the height and depth of a left parenthesis.
  The space increases the consistency of vertical spacing within mathematics in situations where spacing varies because of the characters in the mathematics.

- Custom sets the additional vertical spacing to any value you want.
  You can discard the space if it falls at the end of a page or require that the space be included regardless of where it falls on the page. Required space is useful for leaving room between questions on an examination.

- The vfill fragment adds as much vertical space as can fit on the page.
  If the space falls at the end of the page, the space is removed.

- The rvfill fragment adds as much vertical space as can fit on the page.
  The space is required regardless of where it falls on the page.

---

**Tip**   Place fragments that generate additional vertical space in a paragraph by themselves.

---

▶ **To insert vertical space after a line**

1. Move the insertion point to the line after which you want to add space.

2. From the Insert menu, choose Spacing and then choose Vertical Space (ALT+I, S, V).

3. From the Vertical Space dialog box, choose the vertical space you want or specify a custom space.

4. Choose OK.

   The program inserts the space. When the Invisibles option is turned on, the vertical space appears on the screen as a small green horizontal line; the line doesn't appear in print.

   –or–

1. Move the insertion point to the line after which you want to add space.

2. From the File menu, choose Import Fragment (ALT+F, R).

3. Choose the vertical space fragment you want.

4. Choose OK.

   –or–

1. Move the insertion point to the line after which you want to add space.

2. Click the Fragments popup box.

3. Choose the vertical space fragment you want.

   The program inserts the space. The fragment appears on the screen as a small gray box containing the word *rvfill* or *vfill*.

## Adjusting Page and Line Breaks

Although the typesetting style determines how many lines of text and mathematics fit on a given page and how many characters fit on a given line, you can force a page or line break at a certain point in your document with *SW* fragments. New line commands don't start new paragraphs, but some new page commands do.

▶ **To force a line break**

1. Position the insertion point where you want the line break to occur.

2. From the Insert menu, choose Spacing (ALT+I, S).

3. Choose **Break** (ALT+B).

4. From the **Break** dialog box

- Choose **Newline** (ALT+L) to start a new line at the break.
- Choose **Linebreak** (ALT+K) to start a new line at the break and fully justify the text on the broken line.
- Choose **Custom Newline** (ALT+C) to start a new line at the break.
  The new line will be preceded by the amount of space you specify:
    a. Specify the amount of space to be inserted after the break and before the new line begins.
    b. Specify the unit of measure for the space.
    c. Specify the typeset conditions under which the break can or cannot occur.

5. Choose **OK**.

The program inserts the kind of new line you specified. When the **Invisibles** option is turned on, the new line command appears on the screen as a small vertical line; the line doesn't print. If you specified **Custom Newline**, the length of the line reflects the amount of space you specified.

▶ **To force a page break**

1. Position the insertion point where you want the page break to occur.

2. From the **Insert** menu, choose **Spacing** (ALT+I, S).

3. Choose **Break** (ALT+B).

4. From the **Break** dialog box

- Choose **Newpage** (ALT+P) to start a new page and a new paragraph at the break.
- Choose **Pagebreak** (ALT+G) to start a new page after the line on which the break occurs.

5. Choose **OK**.

The program inserts the page break you specify. When the **Invisibles** option is turned on, the page break appears on the screen as a small horizontal line; the line doesn't appear in print.

## Adding Lines

Although the style you choose determines where lines appear in your document, you may want to add solid lines to set off certain portions of the document. You can specify the thickness and the length of the line and where it falls in relation to the baseline of the surrounding text.

▶ **To add a horizontal line**

1. Position the insertion point where you want the line to begin.

2. From the Insert menu, choose Spacing (ALT+I, S).

3. Choose Rule (ALT+R).

   The system opens the Lift dialog box.

4. In the Lift box, specify the elevation of the line above the baseline.

5. In the Width box, specify the length of the line.

6. In the Height box, specify the height of the line.

7. In the Units box, specify the unit of measure for the specifications.

8. Choose OK.

   For example, if you specify a rule with a lift of 0.0 inch, a width of 1.0 inch, and a height of 0.02 inch, you produce a line like this ——————————. Similarly, if you want a thick, 3-inch line like this ██████████████████████████████ that sits slightly above the baseline, you specify a lift of 0.02 inch, a width of 3.0 inches, and a height of 0.05 inch.

---

**Note**    You can add lines and dotted lines by adding horizontal space. See "Adjusting Horizontal Spacing" earlier in this chapter.

---

# Understanding Formatting

The extensive formatting operations that are so common to working with most word processors are unnecessary in *SW*. The typeface, heading style, page numbering scheme, and all other formatting details are handled automatically by the typesetting styles. For example, styles determine the typeface in which the text and headings are printed. When you choose a style, you choose the point size you want to use for the body of the text. The size of type used for other elements, such as headings and footnotes, changes in proportion to the size of type used in the body. Typesetting styles determine which elements— such as headings or theorem statements—are printed in boldface or italics and which, if any, are underlined. You can add bold or italic emphasis to a word or phrase in the body of your document with text tags, as explained in Chapter 3 "Entering and Editing Text." However, *SW* has no underlining command in the usualsense.

   *SW* typesetting styles determine where headings are located on the page, which document elements, if any, are centered, and whether the page format adjusts for right- and left-hand pages. The styles also determine the page size and orientation, the margins, the number of columns, and the page numbering scheme. See Chapter 11 "Structuring Documents" for information about adding section headings and other structural elements to your document.

Remember that the on-screen appearance and the printed appearance of tagged elements in your document are different. Text that appears indented, in bold, or in italics on the screen may not appear similarly in print. Many document elements, such as columns and headers, don't appear in the main window, but you can see how they will look in print if you preview the document. This section summarizes how *SW* handles the most typical formatting operations and refers you to chapters that contain additional information.

## Boldface

The typesetting style determines which headings are printed in boldface. You can add emphasis to a word or phrase in body text with the **Bold** text tag, if it is present on the list of text tags for the style you choose. Some styles use content-oriented tag names exclusively such as Define or Emphasize instead of appearance-oriented names such as Bold and Italic. See Chapter 3 "Entering and Editing Text."

## Centering

Titles and displayed mathematics are usually centered automatically when printed. The typesetting style determines which document elements are centered. You can use the **Body Center**, **Body Quote**, and **Body Quotation** tags to center a paragraph of body text. See Chapter 3 "Entering and Editing Text."

## Columns

The options available for the typesetting style you choose determine the number of columns printed on a page. If you preview the style, you can examine the column layout.

## Double-Sided Printing

Double-sided printing is a style option that offsets printing differently for odd and even pages. This offset allows for space used when you bind the pages into a book. Usually, book styles use double-sided printing; article and report styles use single-sided printing. Whether your printer actually prints on both sides of the paper depends on the printer itself. See Chapter 5 "Previewing and Printing Documents."

## Duplex Printing

Duplex printing is the term used for printing on both sides of the physical page. Duplex printing is a function of your printer.

## Headers and Footers

The typesetting style determines the presence, layout, and content of headers and footers. You can preview the style to examine the appearance of headers and footers.

## Hyphenation

The program automatically hyphenates your text when you print or preview your document. You can request conditional hyphenation for specific words using the hyphenation fragment, which appears as a dash (–) at the top of the popup list on the Fragments toolbar.

## Indention

The indention of paragraphs, headings, and lists in a document is determined by the style. Remember, indention on the screen does not imply indention in print.

## Italics

The typesetting style determines whether headings, theorem statements, and other document elements are printed in italics. You can use the Italic text tag to add emphasis to a word or phrase that the style does not set in italics. Some styles don't make the Italic tag available—instead, they use tag names such as Emphasize that relate to content rather than appearance. See Chapter 3 "Entering and Editing Text."

## Justification

Justification is determined by the style you choose. Justification is not shown in the main window but you can preview the style or your document to examine the appearance of justified text.

## Kerning

The style you choose determines the kerning, or amount of space between letters and characters. You may occasionally want to use the Spacing command on the Insert menu to add horizontal space to your mathematics or text, or to add space on a line that can be left empty or filled with dots or a line. See "Adjusting Horizontal Space" and "Adjusting Breaks in Mathematical Expressions" earlier in this chapter. See also Chapter 3 "Entering and Editing Text" and Chapter 4 "Entering and Editing Mathematics."

## Leading

Leading is the amount of space between printed lines of text or mathematics. Leading is governed by the style, although you can use the **Spacing** command on the **Insert** menu to add vertical space. See "Adjusting Vertical Space" earlier in this chapter.

## Line Breaks

Although the point of each line break depends upon the style, you can specify line breaks with the **Spacing** command on the **Insert** menu. See "Adjusting Page and Line Breaks" earlier in this chapter.

## Line Spacing

The typesetting style determines the line spacing in your document. Some styles provide a double-spacing option. You can make local adjustments to line spacing using the **Spacing** command on the **Insert** menu. For more information, see the earlier section in this chapter "Adjusting Vertical Spacing."

## Lines and Boxes

You can add lines to your document, as described in "Adding Lines" in this chapter. Also, you can add simple boxes and lines with the options available in the **Decoration** dialog box. See Chapter 4 "Entering and Editing Mathematics."

## Margins

Top, bottom, right, and left margins are determined by the typesetting style. You can see how the margins will look in print if you preview the style or your document.

## Page Breaks

The style determines how many lines of text and mathematics fit on a given page. You may, however, want to use the **Spacing** command on the **Insert** menu to force a page break at a certain point in your document, as described in "Adjusting Page and Line Breaks" in this chapter.

## Page Format

Whether the page format adjusts for right- and left-hand pages depends on the style you choose.

## Page Numbering

Whether the pages of your document are numbered and where those numbers appear on the page are functions of the style.

## Page Orientation

Portrait and landscape page orientations are determined by the printer setup, the typesetting style, and the capabilities of your printer. Preview the document to see how it will appear in print. You can choose landscape printing if you choose to print directly from the previewer.

## Page Size

The size of the printed page depends on the style.

## Tabs

Because tabs are used for visual formatting, *SW* does not use tab stops or respond to tab commands in text. The style you choose automatically controls all indention. The TAB key has no effect on indention, but you can use it to move the insertion point in a table, a matrix, or a dialog box.

## Typeface

The typesetting style determines the typeface in which the text and headings are printed. Some styles provide options for wholesale changes to the typefaces used in the document. Preview the style or your document to see how headings and text will appear in print.

## Type Size

The size of the type used when you print or preview your document is an option of the typesetting style. In most style, you can choose 10- , 11- , or 12-point styles for the body of the text. The size of type used for other elements, including headings and footnotes, changes in proportion to the body type size.

## Underlining

*SW* has no underlining command in the usual sense. The style determines which titles or headings, if any, are underlined. On rare occasions you may want to emphasize a word with underlining instead of with one of the text tags. In those cases you can use the

decoration for Bar Under  in the Decoration dialog box.

# 11 Structuring Documents

The printed appearance of your document is determined by its structure, as interpreted by the typesetting style you choose. At the highest level, document structure consists of *front matter*, *body*, and *back matter*. The front matter includes such parts as a title area or title page, a table of contents, an abstract, a list of figures, and a list of tables. The body of a document can be structured into chapters, sections, subsections, and smaller divisions. The back matter includes such parts as the bibliography, appendixes, and index.

Because different typesetting styles interpret the document structure in different ways, the printed appearance of your document changes when you change the style. Chapter 10 "Applying Styles to Documents" explains how you choose or change your document's style.

## Using Fields to Add Structure

*SW* often uses predefined *fields* to store items that structure your document, such as the information in the front matter of an article or the standard parts of a business letter. A field is displayed with a yellow box that contains the name of the field, followed by the body, or contents, of the field. You add structure to your document by creating, deleting, or editing fields and the information they contain. A field can contain both text and mathematics.

A field can appear in a dialog box or in a document window. For example, when you choose Front Matter from the File menu, *SW* displays the Front Matter dialog box, which contains a field for each front matter item.

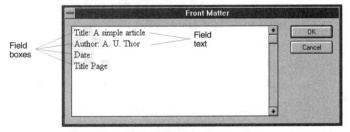

The front matter items, which differ for each typesetting style, are listed on the Item Tag popup list. Typical fields for front matter items are discussed in "Creating and Editing the Front Matter" later in this chapter.

Similarly, when you open or create a letter or memo, *SW* opens a shell document and displays the address, date, and other items in fields in the document window. These fields, which are arranged according to the typesetting style, are discussed in "Structuring Documents with Fields" later in this chapter.

▶ **To complete or edit a field**

- If the field contains predefined information, select it and replace it with the information you want.
  –or–
- If the field is empty, enter the information you want.
  You can use all the editing and entry features available in the *SW* main window.

▶ **To create a continuation paragraph in a field**

1. Position the insertion point at the end of the field.

2. Press ENTER.

   Continuation paragraphs are created automatically when you press ENTER at the end of a paragraph in a field. If you press ENTER accidentally, press BACKSPACE twice to remove the new continuation paragraph.

▶ **To add a new field (or to restore one you have deleted)**

1. Place the insertion point at the end of the paragraph that is to precede the new field.

2. Press ENTER.

3. From the Item Tag popup list, select the field you want.

   The program inserts a yellow box labeled with the field name.

4. Type the text of the field.

   Press ENTER to create additional paragraphs within the field.

▶ **To create two consecutive fields of the same type**

1. Create the first field following the preceding instructions.

2. Place the insertion point at the end of the first field.

3. Press ENTER.

4. Press F2 or click the Remove Item Tag button [⬅•] to end the previous field.

5. From the Item Tag popup list, select the same field tag you used for the preceding field.

   You now have two consecutive yellow field boxes with the same label.

6. Type the text of the second field.

▶ **To delete a field**

1. Select the entire field.

   Click to the right of the end of the preceding field or paragraph and drag down to the end of the last paragraph of the field you want to delete.

2. Choose Delete.

   –or–

1. Delete the text of the field.

2. Press F2 or click the Remove Item Tag button ⬅ to remove the yellow item box.

3. Press DELETE or BACKSPACE to delete the remaining empty paragraph.

▶ **To change the type of a field**

1. Place the insertion point in the field you want to change.

2. From the Item Tag popup list, select the name of the field you want.

   The label on the yellow field box is changed accordingly.

## Creating and Editing the Front Matter

The front matter items available in your document depend on the document type (book, article, letter, etc.) and, in some cases, on the style. The available front matter items include title area or page, table of contents, list of figures, list of tables, and abstract.

Each front matter item defined for the typesetting style is contained in a field in the Front Matter dialog box. You enter the front matter by entering information into each field. The order in which the front matter items are printed is determined by the order in which the fields containing those items appear in the Front Matter dialog box. All text in the Front Matter dialog box must be entered in an appropriate field. Any text not in a field is discarded when you save your document.

▶ **To create or edit your document front matter**

1. From the File menu, choose Front Matter (ALT+F, F).

   The Front Matter dialog box appears with a field for each front matter item. The start of each field is indicated by a yellow box labeled with the name of the field.

2. Enter the front matter information into the various fields.

   You can use all the editing and entry features available in the main window.

3. Edit the existing fields, delete existing fields, or add new fields according to the instructions in "Using Fields to Add Structure" earlier in this chapter.

## Title Area or Title Page

Small documents such as articles are usually printed with the title, date, and author's name and address in a special title area. Larger documents such as books and some reports are printed with that information on a separate title page. The typesetting style determines whether the front matter fields are used to create a title area or a title page and how the front matter information is formatted. Most styles build the title area or title page from the Title, Author, and Date fields.

▶ **To create a title area or title page**

1. From the File menu, choose Front Matter (ALT+F, F).

2. Create or edit the author, title, and date fields.

   Your particular document or style may use other items. These items, if available, are shown on the Item Tag popup list.

3. Place the insertion point at the end of the last field to be used in the title area or page.

4. Press ENTER.

5. From the Item Tag popup list, choose Make Title.

   The Make Title field, a special field with no data, tells *SW* to create the title. If this field is not present, no title is created in the printed document. You can use this feature to turn off the title area or title page temporarily while retaining the information needed to generate it. The Make Title field must appear after all the items used to build the title. The Make Title field should contain no text. *SW* ignores any text you type in this field.

   The order of the Make Title field relative to certain other front matter fields (such as the abstract or Make TOC fields) is also important. In some styles, you can change the order of front matter information by changing the order in which the corresponding fields appear in the Front Matter dialog box.

6. Choose OK.

## Table of Contents

When you print a document, you can instruct *SW* to create a table of contents based on the section tags applied to the headings in the document. The typesetting style you select governs how many heading levels appear in the table of contents.

▶ **To create a table of contents**

1. From the File menu, choose Front Matter (ALT+F, F).

2. Place the insertion point at the end of the field after which the table of contents is to appear.

Remember that the order in which the table of contents, list of figures, and list of tables are printed is determined by the order in which their corresponding fields appear in the **Front Matter** dialog box.

3. Press ENTER.

4. From the Item Tag popup list, choose **Make TOC**.

   The Make TOC field should contain no text. SW ignores any text you type in this field.

5. Choose **OK**.

## List of Figures

Some styles contain a list of figures. When you print a document created with one of these styles, you can instruct *SW* to create a list of figures based on the captions of the graphics in your document. The typesetting style determines the appearance of the list of figures. See Chapter 7 "Using Tables and Graphics" for more information about creating figure captions.

▶ **To create a list of figures**

1. From the **File** menu, choose **Front Matter** (ALT+F, F).

2. Place the insertion point at the end of the field after which the list of figures is to appear.

3. Press ENTER.

4. From the Item Tag popup list, choose **Make LOF**.

   The **Make LOF** field should contain no text. *SW* ignores any text you type in this field.

5. Choose **OK**.

## List of Tables

Some styles contain a list of tables. When you print a document created with one of these styles, you can instruct *SW* to create a list of tables based on the captions of the floating tables in your document. The typesetting style determines the appearance of the list of tables. See Chapter 7 "Using Tables and Graphics" for more information about creating floating tables.

▶ **To create a list of tables**

1. From the **File** menu, choose **Front Matter** (ALT+F, F).

2. Place the insertion point at the end of the field after which the list of tables is to appear.

3. Press ENTER.

4. From the Item Tag popup list, choose Make LOT.

   The Make LOT field should contain no text. *SW* ignores any text you type in this field

5. Choose OK.

## Abstract

▶ **To create an abstract**

1. From the File menu, choose Front Matter (ALT+F, F).

2. If an abstract field is already present, place the cursor immediately to its right and type your abstract.

   If not:

   a. Place the insertion point at the end of the field after which the abstract is to appear.
   b. Press ENTER.
   c. From the Item Tag popup list, choose Abstract.
   d. Type the text of your abstract.

3. Choose OK.

## Structuring the Body of the Document

Section and body paragraph tags define the structure of most *SW* documents. Each paragraph automatically has an associated tag. For most styles, the default is the Body Text tag. You structure the body of your document by applying tags that define chapter, section, and subsection headings, and body paragraph tags that center text elements or set off quotations from the main part of a paragraph. You can provide additional structure by applying item tags to body text paragraphs to indicate statements of theorems, propositions, lemmas, and other theorem-like statements. See Chapter 3 "Entering and Editing Text" and Chapter 4 "Entering and Editing Mathematics" for more information. Certain styles, such as those for letters, memos, and fax messages, have a separate structure that you create using fields for information such as the address, opening, date, and subject. See "Structuring Documents with Fields" later in this chapter.

## Adding Section Headings

*SW* provides tags for chapter, part, and section headings and four levels of subheadings. The tags are in the Section and Body Tag popup list in the Tag toolbar.

When you print your document, *SW* automatically numbers section headings. The typesetting style determines how the section headings appear in print, which heading levels are numbered, the numbering scheme used, and which headings appear in the table of contents. On the screen, section headings often appear larger than body text and in a different color. They also may be indented to help you see the structure of your document.

*SW* is preconfigured with several function key assignments for section and body tags:

| Function Key | Tag |
|---|---|
| F3 | Body Text |
| F11 | Section |
| F12 | Subsection |

Use the Function Keys command on the Tag menu to change the assignment of tags to function keys for faster use.

▶ **To apply a section tag**

1. Place the insertion point within the paragraph to be tagged. If you want to tag several paragraphs, select them.

2. From the Section and Body Tag popup list, click the section tag you want.

   –or–

   Press the function key assigned to the tag.

   –or–

   From the Tag menu, choose Apply (ALT+T, A), select the tag you want, and choose OK.

▶ **To remove a section tag from a paragraph**

• Apply the Body Text tag to the paragraph (F3).

## Adding Theorem and Theorem-Like Statements

In *SW*, you use item tags to identify theorem and theorem-like statements such as propositions, lemmas, and corollaries. The printed appearance of these statements depends on the typesetting style you choose. Many styles use italics for the body of a theorem statement. The Scientific Word User's Guide Style, the style used to print this book, prints theorem statements this way:

**Theorem 1**   *Let S be a sheaf of germs of holomorphic functions. . .*

You can add a label to a theorem by revising the theorem tag; for example,

**Theorem 2 (Fingle)**   $f = nd^3$

Theorem tags are available only in certain styles. If you want to include theorems and theorem-like statements in your document, be sure to preview the style you before you choose it to make sure it includes the elements you need.

### ▶ To enter the statement of a theorem

1. Place the insertion point at the end of the paragraph that is to precede the theorem statement.

2. Press ENTER.

3. From the Item Tag popup list, click **theorem**.

   A yellow box containing the word *Theorem* is displayed at the beginning of the new paragraph.

4. Type the statement of the theorem.

   To create an additional paragraph within the theorem statement:

   a. Press ENTER.
   b. Press BACKSPACE to delete the automatically supplied lead-in.
   c. Type the paragraph.

5. Press ENTER.

6. To end the theorem, click the Remove Item Tag button  or press F2.

### ▶ To apply a theorem tag to one or more existing paragraphs

1. Select the paragraphs.

2. From the Item Tag popup list, click **theorem**.

   A yellow box containing the word *Theorem* is displayed at the beginning of each paragraph in the selection.

   If your selection includes more than one paragraph and you want the paragraphs to be part of a single theorem statement, place the insertion point at the start of each continuation paragraph and press BACKSPACE to delete the yellow box containing the word *Theorem*. Only the first paragraph of the theorem statement should start with one of these boxes unless you want each paragraph to be a separate theorem statement.

### ▶ To remove a theorem tag from a paragraph

- Place the insertion point in the paragraph and click the Remove Item Tag button or press F2.

▶ **To add a theorem label**

1. Open the Lead Item Properties dialog box:

   - Double-click the yellow box containing the word *Theorem*.
     –or–
   - Place the insertion point to the right of the box, then

     - On the Standard toolbar, click the Properties button 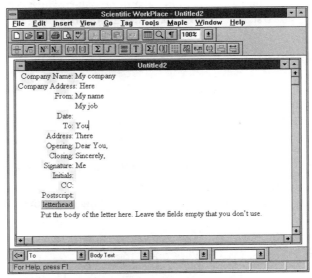 .
       –or–
     - From the Edit menu, choose Properties (ALT+E, O).
       –or–
     - Press CTRL+F5.

2. Select Custom.

3. In the Custom box, enter the label.

4. Choose OK.

Your label is displayed in the yellow box for the theorem. When you print the document, the label appears after the theorem number, as shown in the example above. The choice of brackets (if any) enclosing the label and the font used for the label are determined by the typesetting style.

## Structuring Documents with Fields

The shell documents provided for letters, memos, and fax messages have a number of predefined fields for you to fill in.

In these documents, the on-screen order of the fields above the **Opening** field does not matter because the style determines the actual order of the information when you print or preview.

▶ **To create a letter, memo, or fax message**

1. From the **File** menu, choose **New** (ALT+F, N).

2. Select **Letter**, **Memo**, or **FAX** in the **Document Type** box.

3. Select a style from the **Style** box.

4. Choose **OK** to load a shell document with a number of predefined fields.

5. Replace the text in the fields with your own information.

▶ **To edit or delete fields**

- Follow the instructions for editing fields in "Using Fields to Add Structure" at the beginning of this chapter.

---

**Important**   If a letter or memo style has an Opening field, you must not delete it—it is used as a signal to create the letter or document when you print or preview.

---

## Creating the Back Matter

Your document may contain supplementary material such as appendixes, a bibliography, or an index. Unlike the front matter, which *SW* creates for you from information you supply in the **Front Matter** dialog box, you create the back matter of your document in the *SW* main window. The typesetting style you choose determines the printed appearance of back matter.

## Appendixes

You enter the content of an appendix exactly as you enter the content of a chapter or section in the body of your document. You can number appendixes automatically.

▶ **To create appendixes that are numbered automatically**

1. Move the insertion point to the end of the line that is to precede the first appendix.

2. Press ENTER.

3. Insert the **appendix** fragment.

   This simple fragment is a TEX field containing \appendix.

4. Type the appendixes.

- If you're using a book style, tag the heading of each appendix as a chapter head.
- If you're using an article style, tag the heading of each appendix as a section head.

*SW* does not display the appendix numbers on the screen. In print, the numbering scheme is determined by the typesetting style. Typically, appendixes are numbered sequentially as A, B, C, etc., and appendix subheadings are numbered sequentially beginning with the letter of the appendix, as in A.1, A.2, A.3, B.1, B.2, B.3, etc.

## Bibliographies

Bibliographies, or references, are lists of articles, books, or other sources that appear at the end of a document or chapter. With *SW,* you can create bibliographies manually or you can use a special program called BIBTEXto create bibliographies automatically.

- **Manual Bibliographies.** A manual bibliography, which works just like a series of cross-references, is convenient when the list of references is short and you do not plan to use those references in other articles or books. You must format entries yourself, a disadvantage if you decide to change bibliography styles. For example, one journal might require book titles in italics and another might require those same titles in boldface.

- BIBTEX **Bibliographies.** BIBTEX is a public domain program created by Oren Patashnik. The program creates a bibliography by extracting references from a database using the citations you insert into an *SW* document. A bibliography created with BIBTEX can be changed from one style to another easily. To use BIBTEX, you must first create a separate database of bibliographic entries. This database can be used as often as you like in other documents. You can use BIBTEX databases from other sources. The main disadvantage of using BIBTEX to create bibliographies is the effort required to set up a database in the first place.

### Creating a Bibliography Manually

Creating a manual bibliography involves three steps: specifying a manual bibliography; creating a list of *bibliography items*; and creating a series of *citations*, or references, to those items. Generally, the list of bibliography items appears at the end of your document. The printed format of the items in the list is determined by the typesetting style. Many styles automatically number each item in the bibliography list, but you can replace a number with a label that appears in print, both in the bibliography list and everywhere you place a citation for the labeled bibliography item.

▶ **To specify a manual bibliography**

1. From the File menu, choose Document Info (ALT+F, D).

2. Choose the Bib Choice tab.

3. Select Manual Entry.

4. Choose OK.

▶ **To create a list of bibliography items**

1. Move the insertion point to the end of the line that is to precede the bibliography, typically the end of the document.

2. Press ENTER.

3. From the Item Tag popup list, click **Bibliography item**.

   The program displays the **Bibliography Item Properties** dialog box.

4. Enter a unique key for the item.

   You use this name when you create a citation for the bibliography item. Click the arrow next to the **Key** box to display a list of keys already in use in your document, or press DOWN ARROW to scroll through the list.

5. If you want a label for the item instead of the number automatically generated by the style, enter the label in the **Label** box.

   The label can include mathematics.

6. Choose **OK**.

7. Type the bibliographic information for the item.

8. If you want another bibliography item, press ENTER and repeat steps 4–7.

9. At the end of the bibliography, press ENTER to complete the last item.

10. To complete the bibliography item list , click the Remove Item Tag button  or press F2.

On the screen, the key for each item appears in a yellow box next to the bibliographic information. In print, the keys do not appear and the items are numbered sequentially unless you added labels.

▶ **To create a citation for a manually created bibliography item**

1. Place the insertion point where you want the citation to appear.

2. On the Field toolbar, click the Citation button.

   –or–

   From the **Insert** menu, choose **Field** and then choose **Citation** (ALT+I, E, C).

   The program displays the **Citation** dialog box:

3. Enter the key for the bibliographic item you want to cite.

Click the arrow next to the **Key** box to display a list of all keys defined for the document.

4. If you want to add a comment to the citation, enter it in the **Remark** box.

You can enter both text and math in the comment.

5. Choose **OK**.

*SW* displays the citation on the screen in a small gray box containing the word *cite* and the key you entered, like this:

$$\ldots \text{as noted in an earlier article } \boxed{\text{cite: example}}.$$

When you print the document, *SW* creates the correct reference by substituting the number of the keyed bibliography item in place of the key:

$$\ldots \text{as noted in an earlier article } [1].$$

If you also entered a remark, *SW* prints it after the number:

$$\ldots \text{as noted in an earlier article } [1, \text{based on Fermat's principle}].$$

The style you choose for your document determines the appearance of the citation in print.

### Creating a Bibliography Using BiBTeX

Creating a bibliography with BIBTEX requires that you first create or obtain a database of bibliographic items in BIBTEX form, then refer to the items with citations in your document. A BIBTEX database is an ASCII file you create and edit with a standard ASCII editor. Each record in the database contains the information necessary to create a bibliographic entry in a document. The records are stored in a logical format with no visual (formatting) information. *SW* provides several sample databases. The databases have a `.bib` file extension and are located in the `bibtex` subdirectory of your *SW* directory.

The `bibtex` directory contains the BIBTEX program and a number of additional files. The directory contains a `readme` file that contains more information. You can load this file into *SW* using the ASCII filter, then print it to examine it. The files `btx-doc.tex` and `btxhak.tex` are the original documentation provided by Oren Patashnik.

When BIBTEX creates a bibliographic entry from a database record, it does so with the aid of a BIBTEX style file that gives instructions on how to format the entry for a particular journal or book. Files with the extension .bst are BIBTEX style files. The advantage of this approach is that you can change the format of your citations and bibliographic entries simply by changing the BIBTEX style. Instructions for creating and editing BIBTEX databases are in the read.me file in the bibtex directory.

Once you have established one or more BIBTEX databases containing the references you want, using the databases to create a bibliography in your document is much more convenient than using the manual process described earlier. You create the citations and BIBTEX generates the bibliography automatically. If your database of bibliography entries is very large, finding the reference you want to cite can be time-consuming. You can narrow your search by specifying search criteria, or *key filters*, for the database items, then choosing from among those that fit the filters.

Creating a BIBTEX bibliography involves four steps: specifying a BIBTEX bibliography, creating a series of citations in the body of your document, inserting an instruction to include the bibliography, and generating the bibliography.

### ▶ To specify a BibTEX bibliography

1. From the File menu, choose Document Info (ALT+F, D).

2. Choose the Bib Choice tab sheet.

3. Select BibTeX.

4. Choose OK.

Choosing BIBTEX for your bibliography changes the Citation dialog box so you can select entries from a BIBTEX database.

### ▶ To create a citation for an item in a BIBTEX database

1. Position the insertion point in the body of your document where you want the citation to appear.

2. On the Field toolbar, click the Citation button 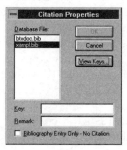 .

   −or−

   From the Insert menu, choose Field and then choose Citation (ALT+I, E, C).

   The program opens the Citation dialog box:

3. If you know the key for the database item you want to cite, enter it in the **Key** box.

Otherwise, select the key for the item directly from a BIBTEX database:

    a. Select one of the databases listed.

    b. Select **View Keys**.

       This displays a dialog box showing the keys for database items that satisfy the current search criteria:

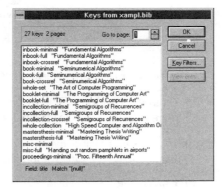

    c. Page up or down through the list using the **Go to page** box to select the item you want.

    d. If you want to view the full text of the item, choose **View entry**.

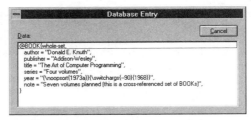

    e. When you find the item you want, select it and choose **OK**.

4. Choose **OK**.

The program inserts a citation for the item whose key you selected.

▶ **To change the search criteria**

1. On the Field toolbar, click the Citation button 　.

   –or–

From the **Insert** menu, choose **Field** and then choose **Citation** (ALT+I, E, C).

The program opens the **Citation** dialog box.

2. Select one of the databases listed.

3. Select **View Keys**.

The program opens the **Key Filters** dialog box:

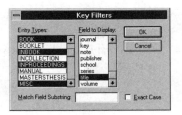

4. From the **Entry Types** list, choose the types of items you want to cite.

5. From the **Fields to Display** list, choose the bibliography entry fields you want to display in the **Keys from *database*.**bib dialog box.

6. If you want to show every item in the database, select all items in the **Entry Types** list and make sure the **Match Field Substring** box is empty.

7. Choose **OK**.

In the **Keys from *database*.**bib dialog box, the program lists the items in the selected database that match the search criteria you specified.

▶ **To insert an instruction to include the bibliography**

1. Move the insertion point to the end of the line that is to precede the bibliography, typically the end of the document.

2. Press ENTER.

3. From the **Insert** menu, choose **Field** and then choose **Bibliography** (ALT+I, E, B).

4. From the **Bibliography Properties** dialog box, select the BIBTEX database (or databases) containing your references.

5. Select a BIBTEX style for formatting those references.

For more information about the various styles, preview the `btxdoc.tex` file in the `bibtex` directory.

6. Choose **OK**.

▶ **To generate the bibliography**

1. Save your document.

2. From the **File** menu, choose **Compile** (ALT+F, L).

The **Compile** option is available only if you have made no changes to your document since it was last saved.

3. Check **Generate a Bibliography**.

The **Generate a Bibliography** option is available only if you have inserted a BibTEX bibliography field in your document.

4. Choose **OK**.

The program processes your document first through LATEXand then through BibTEX. The BibTEX stage automatically extracts and formats references from the database using the citations in the body of the document. The references are placed in a file with the same name as your document but with the extension `.bbl`. When you next print or preview your document, this file is automatically included at the point where you inserted the instruction to include the bibliography.

Whenever you add or delete citations in the body of your document, you must regenerate the bibliography before you preview or print.

## Indexes

In *SW* you can generate an index to help your readers find information in your document. The index to your document can have primary, secondary, and tertiary entries, as well as cross-references. The appearance of the index is determined by the typesetting style you choose for your document. Generating an index involves several steps:

1. Creating an index entry for each item to be indexed.

2. Inserting a command to include the index.

3. Generating the index.

### Creating Index Entries

Each time you want a reference to an item to appear in the index, you must create an index entry. The index entry indicates the *primary entry* under which you want the item to appear in the index and the *secondary* and *tertiary entries,* if any. For example, in the sentence

> You can add emphasis to text selections by applying tags such as bold, italics, or small caps.

you might index *bold* under its own primary entry and also as a secondary entry under the primary entries of *emphasis*, *text appearance*, and *text tags*. The index entries would appear as follows:

bold, 7

⋮

emphasis
bold, 7

⋮

text appearance
bold, 7

⋮

text tags
bold, 7

Within the index, references appear in alphabetical order, with symbols preceding words. You can specify a special appearance for an index entry so that a symbol can appear in the correct alphabetical order in the index. This example contains the symbol { }, which is a nonalphabetic item:

Curly braces, { }, are just one of many enclosures available.

You might use the words *curly braces* to index the symbol under its own primary entry and to create a secondary entry under the primary entries *brackets* and *enclosures*. If you also specified that each entry should appear as the braces themselves instead of the words, the index entries would appear as follows:

brackets
{ }, 34

⋮

{ }, 34

⋮

enclosures
{ }, 34

Cross-references within the index can help your reader find information. Use cross-references to point to other entries in the index. For example, if all entries pertaining to *Scientific WorkPlace* are indexed under *SW*, you might help the reader find the entries by creating a cross-reference to *SW*. Your index would then have an entry like this:

Scientific WorkPlace
*See* SW

▶ **To create an index entry**

1. Position the insertion point to the right of the item you want to index.

2. On the Field toolbar, click the Index Entry button  .

   –or–

   From the **Insert** menu, choose **Field** and then choose **Index Entry** (ALT+I, E, I).

The program opens the **Index Entry** dialog box:

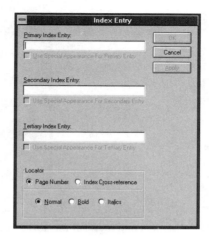

3. If the entry is a primary index entry, fill in the **Primary Index Entry** box with the index entry for the item.

   If you want to specify a special appearance for the index entry, check **Use Special Appearance for Primary Entry** and enter the index entry exactly as you want it to appear. You can use the toolbars and commands in the $SW$ main window to create a special appearance for an index entry.

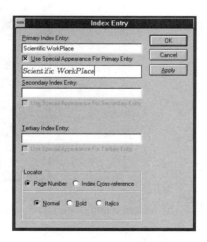

4. If the entry is a secondary index entry, fill the **Primary Index Entry** box with the primary index entry and then fill the **Secondary Index Entry** box with the secondary index entry for the item.

   You can enter a special appearance for the secondary entry.

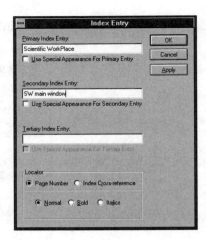

5. If the entry is a tertiary index entry, enter the primary and secondary index entries for the item and then fill the **Tertiary Index Entry** box with the tertiary index entry.

6. If you want to change the appearance of the page number for the index entry, from the **Locator** area choose **Bold** or **Italic**.

7. If you want to enter cross-references, from the **Locator** area check **Index Cross-reference** and then enter the references in the **Cross-reference(s)** box.

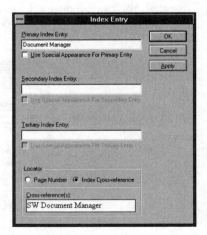

The program automatically supplies the word *See*.

8. Choose **OK**.

### Adding the Index to Your Document

Your document must contain a command to include the index file in your document. The command is available as an *SW* fragment.

▶ **To include the index in your document**

1. Place the insertion point at the position in the document where you want the index to appear.

2. From the File menu, choose Import Fragment (ALT+F, R).

3. Choose the index fragment.

4. Choose OK.

   –or–

1. Place the insertion point at the position in the document where you want the index to appear.

2. From the Fragments popup list, click index.

### Creating a LaTeX Index File

*SW* uses the MakeIndex program to build your index from the index entries you create. MakeIndex creates the index as a sorted LaTeX file with the same name as your document and an .ind extension.

▶ **To create an index file**

1. Save your document.

2. From the File menu, choose Compile (ALT+F, M).

3. Check Generate an Index.

   The program automatically sets the number of LaTeX passes to 2. Multiple passes are needed to resolve all cross-references.

4. Choose OK.

   Preview or print your document to examine the index.

## Document Information

For each document you create, *SW* maintains a body of information called the Document Info. The Document Info maintains general information about your document, including when it was created and how it is stored, and provides a direct way to insert LaTeX commands before the front matter. The Document Info dialog box contains these tab sheets:

- The General tab sheet contains the file name and directory location, the date the document was created and most recently revised, the language used to create the document, and the name of the typist. The tab sheet also contains the control that saves the document for quick loading. See Chapter 2 "Opening and Closing Documents" for more information about quick-load documents.

- The Comments tab sheet stores any notes about the document that you want to preserve. You can have the explanation appear at the beginning of the LaTeX document and in the tab sheet. Comments are not printed, but you can use Document Info to view them.

- The Bib Choice tab sheet contains a toggle to create a manual bibliography or a BibTeX bibliography.

- The Preamble tab sheet contains LaTeX commands that you want to insert at the beginning of your document.

- The File Content tab sheet contains commands that you enter to view and modify any LaTeX 2$_\varepsilon$ imbedded documents.

---

**Important**   Be sure you are very familiar with LaTeX before you make entries in the Preamble or File Content tab sheets. Damage to your document can result from incorrect LaTeX.

---

▶ **To view or change the Document Info**

1. From the File menu, choose Document Info (ALT+F, D).

2. Select the tab sheet you want from the Document Info dialog box.

3. Enter the information you want in the tab sheet.

4. Choose OK.

# 12 Managing Documents

You can use *SW* document management techniques to simplify working with *SW* documents and their associated files. You can manage large documents such as books more easily using quick-load techniques or using master documents and subdocuments. And with the *SW* Document Manager you can handle subdocuments and graphics correctly when you copy, delete, or rename a document. Also, the *SW* Document Manager helps you clean up the auxiliary files generated by the preview and print processes and simplifies working with files sent or received by electronic mail or on disk.

## Managing Large Documents

In *SW* you can speed working with large documents by saving them for quick loading. Chapter 2 "Opening and Closing Documents" contains more information about quick-load files.

You can also break large documents into smaller, more manageable pieces called *subdocuments* and then include all the subdocuments in a *master document.* Subdocuments, which can be used in any number of master documents, do not contain their own document information. Instead, they use the style, Document Info, and front matter defined for the master document. When you compile or preview a subdocument, *SW* resolves only those cross-references that are internal to the subdocument.

A master document differs from a standard *SW* document only in that it contains a field for including each subdocument. The master document stores the style, Document Info, and front matter defined for the entire large document. Any changes you make to the style, Document Info, and front matter of the master document affect all the subdocuments in exactly the same way. When you compile and preview a master document, *SW* resolves the cross-references in all subdocuments and generates the table of contents and the index, if specified, for the whole document.

The manual you are now reading was created using an *SW* master document that includes the chapters and appendixes as subdocuments. The chapters and appendixes were typed, corrected, and printed separately. The master document was used to create the table of contents, the index, and cross-references. Drafts were printed on a laser printer, and the final copy in the form of negatives was output directly from PostScript files.

Creating a large document involves three steps: creating the master document, creating the subdocuments from within the master document, then opening each subdocument and typing your text.

## Creating a Master Document

Just like you create any *SW* document, you can build a master document from a new file, a shell document, or a copy of any other *SW* document. Master documents are ordinarily quite small, often consisting of little more than a list of the fields that cause subdocuments to be included. You can edit the master document at any time to choose the typesetting style for the document, specify the front or back matter, include sub-documents, or type information you want as part of the master document. *SW* applies the document information of the master document—the style, Document Info, and front matter—to all the subdocuments. The changes you make in the master document to the document information affect all the subdocuments. If the subdocuments are open, they must be saved and reloaded. If the master document is a quick-load file, it must be exported to TEX before the subdocuments will be affected. Refer to Chapter 10 "Applying Styles to Documents" for more information about choosing a typesetting style.

When you create a large document, we suggest you set up a separate directory to hold it. The subdocuments can reside in the same directory as the master document or in subdirectories of that directory.

▶ **To create a master document**

1. Open a new file:

   - On the Standard toolbar, click the New button ⬜ .
     –or–
   - From the **File** menu, choose **New** (ALT+F, N).

2. From the **File** menu, choose **Style** (ALT+F, Y) to select the typesetting style for the document.

   The program applies the style you select for the master document to the entire document. Refer to Chapter 10 "Applying Styles to Documents" for more information.

3. From the **File** menu, choose **Document Info** (ALT+F, D) to enter information about the document, then choose **OK**.

4. From the **File** menu, choose **Front Matter** (ALT+F, F) to set up the front matter, then choose **OK**.

   For more information about document information and front matter, see Chapter 11 "Structuring Documents."

5. Enter any text you want as part of the master document.

6. From the **File** menu, choose **Save** (ALT+F, S) to name and save the master document.

   Save the document in the directory that is to contain the master document, all the subdocuments, and any subdocument directories.

7. Create the empty subdocuments.

See "Creating a Subdocument" for further instructions.

It is this step alone that makes a master document different from a standard *SW* document.

8. From the **File** menu, choose **Save** (ALT+F, S) to save the master document again.

## Creating a Subdocument

Each subdocument is a separate document that you can edit independently of the master document. You create a subdocument in two steps. First, you insert an empty subdocument in the master document. Second, you open the subdocument and type its content. You can use an existing document as a subdocument, in which case both the original document and the subdocument share any associated graphics. You can use a subdocument in several master documents. Whenever you open a subdocument, *SW* uses the Document Info, front matter, and style of its most recent master document.

---

**Note**    Subdocuments are different from standard documents because they do not have their own style nor contain any of their own document information. Unlike standard documents, subdocuments must be previewed from within *SW* rather than with the TrueTEX icon in the *SW* program group.

---

▶ **To insert a subdocument into a master document**

1. Open the master document.

2. Move the insertion point where you want to include the subdocument.

3. On the Field toolbar, click the Subdocument button  .

–or–

From the **Insert** menu, choose **Field** and then choose **Subdocument** (ALT+I, E, S).

If the **Subdocument** menu item is dimmed or the **Subdocument** dialog box doesn't open, you must name and save your master document before you can proceed.

4. If you want to choose a directory other than the master document directory for the subdocument, click the box next to the **Subdocument Directory** area.

5. Choose **OK** to close the **Subdocument Directory** dialog box.

6. In the **Subdocument** area, enter a name for the subdocument.

Click the arrow next to the entry box to display a list of all documents in the directory you selected.

7. Choose OK.

When you choose OK, the program inserts a field in the master document that includes the subdocument whenever the master document is previewed or printed. If you do not supply the name of an existing subdocument, a new empty subdocument is created using the name you give.

If the subdocument has been included previously in another master document, the program displays a message asking you to confirm the inclusion of the subdocument.

- Choose Yes to include the subdocument.
- Choose No to return to the Subdocument dialog box and specify a different subdocument.

8. Repeat steps 2–7 for each subdocument you want to include.

You can place any text you want before, between, or after the fields that include the subdocuments.

9. From the File menu, choose Save (ALT+F, S) to save the master document.

---

**Note**   You can use the same steps to create a subdocument of a subdocument. We recommend this only in the case of very long documents.

---

▶ **To make an existing document a subdocument**

1. Open the master document.

2. Insert an empty subdocument using a different name from the existing document.

3. Choose OK.

4. Save the master document.

5. Open the new subdocument.

6. From the File menu, choose Import Contents (ALT+F, T).

7. Select the existing document you want to use as the subdocument.

8. Choose OK to copy the contents of the document into the new subdocument.

9. Save the subdocument.

The original document is unchanged. Use the Document Manager command on the File menu to delete the original document if you no longer need it as a separate document.

**▶ To type or edit a subdocument**

1. On the Standard toolbar, click the Open button  .

   –or–

   From the **File** menu, choose **Open** (ALT+F, O).

2. Select the subdocument you want to type or edit.

3. Choose **OK** to open the subdocument.

   The Title bar shows the name of the master document and the name of the subdocument.

4. Type text or make editing changes as you would for any other *SW* document.

5. From the **File** menu, choose **Save** (ALT+F, S) to save the subdocument.

## Previewing and Printing Subdocuments and Documents

When you preview or print subdocuments individually, *SW* resolves any cross-references that are internal to the subdocument but not cross-references to other subdocuments or to the master document. To resolve all the cross-references, compile the master document. To print your entire document with all cross-references resolved, print the master document.

When you print a master document, you can suppress the printing of individual subdocuments. As the DVI file for the master document is compiled, all cross-references in the master document are resolved, even for suppressed subdocuments. By suppressing all subdocuments but one, for example, you can print a single subdocument with all the cross-references correctly resolved.

**▶ To preview or print a subdocument**

1. Open the subdocument.

2. From the **File** menu, choose **Preview** (ALT+F, W) or **Print** (ALT+F, P).

   The program begins to compile your document with LATEX and displays the **Counters** dialog box.

3. If you want to change the initialization point of the chapter or page numbers, click the corresponding arrows to change the value.

4. If you want to change the initialization point of any other numbers, such as equation, section, or table numbers:

   a. Choose **Other Counters**.
   b. In the **Counter** box, select the counter for which you want a different initialization point.

      c. In the Value box, set the initialization point you want.

      d. Choose OK.

5. Choose OK to leave the Counters dialog box.

The program processes the document and then displays it in the previewer or prints it, according to your instructions. See Chapter 5 "Previewing and Printing Documents" for more information about the LATEX process.

### ▶ To compile, preview, or print an entire document

1. Open the master document.

2. From the File menu, choose Compile (ALT+F, L), Preview (ALT+F, W), or Print (ALT+F, P).

The program compiles the entire document, resolving all cross-references and creating the table of contents and index.

### ▶ To suppress the previewing or printing of subdocuments in a master document

1. Open the master document.

2. Select the field that includes the subdocument you do not want to preview or print and choose Properties.

    –or–

Double-click the field.

3. Check the Suppress Printing box.

4. Choose OK.

The field for the subdocument now contains the word *Exclude* instead of *Include*.

5. Repeat steps 2–4 for each document to be suppressed.

6. Preview or print the master document.

The program compiles the entire document, resolving all cross-references in all subdocuments but printing only that part of the entire document that was not suppressed.

---

**Note**  If you save the master document with subdocuments suppressed, they remain suppressed until you change the Suppress Printing box for each one.

---

## Making a Subdocument into a Document

You can copy an *SW* subdocument into a separate stand-alone document that has all the characteristics and document information of the master document. Once the subdocument has been copied, you can delete it with the Document Manager command if it is no longer used by the master document.

▶ **To make an existing subdocument into a separate stand-alone document**

1. Open the master document.

2. From the File menu, choose Save As (ALT+F, A).

3. Enter the name of the new document.

4. Choose OK.

   The program saves the new document with all the document characteristics defined for the original master document. The style, Document Info, and front matter sections are copied exactly.

5. Delete the contents of the body of the new document.

   The document now has a defined environment but no content.

6. From the File menu, choose Import Contents (ALT+F, T).

7. Select the subdocument you want to copy.

8. Choose OK to copy the content of the subdocument into the new document.

9. From the File menu, choose Save (ALT+F, S).

## Using the SW Document Manager

An *SW* document is associated with many files in addition to the one that contains the document itself. Some of these files may contain subdocuments, graphics, or style information. Others, such as index files and tables of contents, are generated by *SW* when you compile, preview, or print the document. *SW* retains some of these generated files for use the next time you preview or print the document; *SW* retains others for use in maintaining information about program operation or for diagnosing problems. The list that follows describes the types of files most commonly used by *SW*. Depending on the file type, *SW* may not store the associated files in the same directory that holds the document file. When you need to copy an *SW* document, send it to another location, delete it, or rename it, you must make sure that all files associated with your document are treated correctly.

The *SW* Document Manager simplifies working with *SW* documents and their associated files. With the *SW* Document Manager, you can correctly copy and rename documents and delete those files you no longer need, especially generated files. You can properly delete documents that contain subdocuments and references to graphics files. Additionally, with the *SW* Document Manager you can *wrap* a document and its associated files so it can be sent by electronic mail or on disk, and you can *unwrap* a document that has been wrapped. To ensure that all files associated with your *SW* documents are treated correctly, use the *SW* Document Manager to perform all *SW* document management operations.

These are the types of files most commonly used by *SW*:

| Extension | File Type |
|---|---|
| `.aaa....aaz` | Temporary file format of an open file |
| `.aut` | Auto Save file |
| `.aux` | File generated by LaTeX for cross-references, etc. |
| `.bak` | Backup of a `.tex` file |
| `.bbl` | Bibliography file created by BibTeX |
| `.bib` | BibTeX bibliography database |
| `.blg` | Bibliography log file generated by BibTeX |
| `.bst` | Bibliography style file |
| `.cdx` | *SW* Quick Load document file |
| `.cfg` | LaTeX $2_\varepsilon$ document class configuration file |
| `.clo` | LaTeX $2_\varepsilon$ document class option file |
| `.clm` | Dictionary containing user-supplied words |
| `.cls` | LaTeX $2_\varepsilon$ document class file |
| `.clx` | RAM dictionary |
| `.cst` | File controlling the screen appearance of documents |
| `.def` | LaTeX $2_\varepsilon$ definition file |
| `.dll` | Dynamic-link library file |
| `.drv` | LaTeX $2_\varepsilon$ driver file |
| `.dvi` | Device-independent file generated by TeX and required to print or preview |
| `.env` | Spelling environment file |
| `.exe` | Executable file |
| `.fd` | LaTeX font definition |
| `.flt` | Image Stream graphics filter |
| `.fmt` | TeX format file |
| `.glo` | Glossary files |
| `.gls` | Glossary for inclusion in a document |
| `.hlp` | Online help file |
| `.idx` | Index file created by LaTeX, input for MakeIndex |
| `.ilg` | Index log file generated by MakeIndex |
| `.ind` | Index file created by MakeIndex for inclusion in a document |
| `.ini` | Initial settings file |
| `.ist` | MakeIndex index style file |
| `.lat` | List of predefined styles and the files associated with them |
| `.lex` | Unabridged dictionary |
| `.lis` | Report file generated by LaTeX, alternative to `.log` file |
| `.lof` | List of figures generated by LaTeX |
| `.log` | Report file generated by LaTeX |
| `.lot` | List of tables generated by LaTeX |
| `.ltx` | LaTeX $2_\varepsilon$ input file for initex |
| `.m` | Maple program file |
| `.shl` | *SW* shell document |
| `.sty` | LaTeX style file |
| `.tci` | *SW* style organizer file, `styles.tci` |
| `.tex` | *SW* document, TeX or LaTeX file |
| `.toc` | Table of contents generated by LaTeX |
| `.ttf` | TrueType font file |

### ▶ To start the SW Document Manager

- Click the SW Document Manager icon in the *SW* program group    SW Document . Manager

    –or–

- From the **File** menu, choose **Document Manager** (ALT+F, M).
  The *SW* Document Manager window opens:

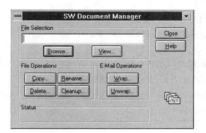

Use the window to select an *SW* document and to select the operation you want the *SW* Document Manager to perform on the document. You can also use the window to view the first part of the selected document.

### ▶ To specify a document

- In the **File Selection** box, enter the complete path name of the document file.
  The complete path name includes the drive, directory, and name of the document file.

    –or–

- Press **Browse** to display the **Open** dialog box, then use the dialog box to select the drive, directory, and name of the document file.

### ▶ To view the first part of the selected file

- Choose **View** from the *SW* Document Manager window.

### ▶ To choose a Document Manager operation

- Click the button for the operation you want.
    –or–
- Press TAB to move to the button for the operation you want, then press ENTER or SPACEBAR.

| Operation | Use |
|---|---|
| Copy | Copy a document and all its associated files to a new location. |
| Delete | Delete a document and all associated auxiliary, graphic, and sub-document files. |
| Rename | Rename a document, subdocument, or graphic file and correct all references to the renamed file. |
| Cleanup | Delete auxiliary and generated files that are no longer needed. |
| Wrap | Gather in a single text file all files that accompany a document file. |
| Unwrap | Break a wrapped file into separate files again. |

The program opens a tabbed dialog box associated with the operation you chose. The *SW* Document Manager focuses on *SW* files and their current or destination location. If you chose to wrap or unwrap a document, the tab sheets also reflect the location of the associated style files.

---

**Important**   When you choose OK to accept the settings in any tab sheet, you accept the settings in *all* tab sheets in the dialog box. Similarly, when you choose Cancel to discard the settings in any tab sheet, you discard the settings in *all* tab sheets in the dialog box.

---

## Directory and File Lists

Each *SW* Document Manager tab sheet has two panels, one indicating the drive and directory location of the files needed for the selected operation and the other indicating the files themselves. Subdirectories are indicated with indention. The panels on the Document, Graphics, and Style tab sheets operate the same way for all Document Manager operations.

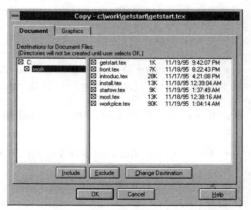

When you select a directory from the panel on the left, the panel on the right lists any files in that directory that are associated with the document you specified for the *SW* Document Manager operation. The file names are listed along with their size, date, and time of last modification.

## Including and Excluding Files

By default, all possible files associated with the selected document are included in each *SW* Document Manager operation. The check boxes in the directory and file panels reflect whether files are included or excluded from the operation: The check boxes in the file panel reflect the selection status of the files themselves. The check boxes in the directory panel reflect the selection status of all the files in the directory. A checked box next to a directory name indicates that all files in the directory will be included in the operation. A partially checked box, used only in the directory panel, indicates that some of the files in the directory and its subdirectories will be included. An unchecked box indicates that none of the files in the directory will be included; that is, they will all be excluded from the operation.

| Description | | Directories | Files |
|---|---|---|---|
| ☒ | Checked | All files in the directory are included in the operation | File is included in the operation |
| ☒ | Partially checked | Some but not all files in sub-directories are included in the operation | |
| ☐ | Unchecked | No files in the directory are included in the operation | File is excluded from the operation |

You may want to exclude certain files from your *SW* Document Manager operations, especially if you are wrapping files to send to someone who also has an *SW* installation. For example, snapshots of Maple plots can be very large. If you are sending a file containing snapshots to someone with a *Scientific WorkPlace* installation, the plots can be recreated in that installation. You may also want to exclude standard *SW* styles.

You can exclude files from the operation on a file-by-file basis or you can exclude all files in a selected directory. Including and excluding files does not apply to the Rename operation.

▶ **To exclude a file from an SW Document Manager operation**

- Uncheck the check box next to the file name in the panel on the right.

  –or–

- Select the file and choose Exclude (ALT+E).

▶ **To exclude multiple files at once**

- Uncheck the check box next to the file names in the panel on the right.

  –or–

1. Hold down the CTRL key while you select multiple files.

2. Choose Exclude (ALT+E).

  –or–

1. Select the first file in a range.

2. Press the SHIFT key.

3. Select the last file in the range.

4. Choose **Exclude** (ALT+E).

### ▶ To exclude all files in a directory

- Uncheck the check box next to the directory name in the panel on the left.

  —or—

- Select the directory and choose **Exclude** (ALT+E).

### ▶ To include all files in a directory

- Check the check box next to the directory name in the panel on the left.

  —or—

- Select the directory and choose **Include** (ALT+I).

### Working with Graphics Files

The dialog box associated with the **Copy, Delete, Wrap,** and **Unwrap** operations has a **Graphics** tab sheet. Just like the **Document** tab sheet, the **Graphics** tab sheet lists the directories and files of graphics associated with the document file. You can exclude them just as you exclude document files.

Note that clipboard pasted pictures that have not been given a name and plot snapshots are stored in *SW* files with names that are meaningful only to the program. These file names are *not* displayed in the file list. You can exclude them from the *SW* Document Manager operation by clearing the check boxes next to **Clipboard Pasted Pictures** and **Plot Snapshots**.

The box labeled **Exclude All Graphics** overrides any other settings on the tab sheet and disables all the controls. If the box is checked, no graphics files will be included in the operation.

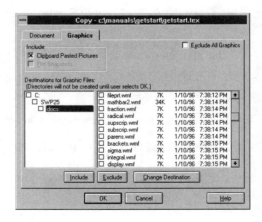

### Working with Style Files

The dialog box associated with the **Wrap** and **Unwrap** operations has a **Styles** tab sheet. Just like the **Document** and **Graphics** tab sheets, the **Styles** tab sheet lists the directories and files containing the typesetting styles necessary to compile the document with LaTeX.

The box labeled **Exclude All Style Files** overrides any other settings on the tab sheet and disables all the controls. If the box is checked, no style files will be included in the operation.

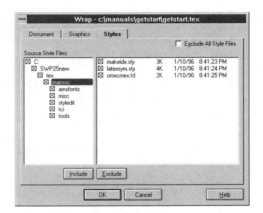

### Changing Destination Directories

When you copy or unwrap a document, you can choose the destination of the document and its associated files. The *SW* Document Manager then restructures the destination directories of all associated files so that all graphic and subdocument files are located in the same directory as the document or in subdirectories within that directory. If you designate a new directory, the *SW* Document Manager will create it when you choose **OK** to begin the copy or unwrap operation.

---

**Important**   The *SW* Document Manager does not actually create directories or copy files until you choose **OK**.

---

▶ **To change a file or directory destination**

1. From the files panel in the **Copy** or **Unwrap** dialog box, select the file or directory for which you want to change the destination.

2. Choose **Change Destination** (ALT+C).

   The program opens the **Move Files** or **Move Directory** dialog box:

3. The default file destination is displayed in the **Folders** box.

4. If you want to choose an existing directory as the destination for the file, double-click its name in the directory list.

5. If you want to create a new directory, select an existing directory to contain the new directory and then choose **Create**.

   The system opens the **Create Directory** dialog box.

   a. Enter a name for the new directory.
   b. Choose **OK**.

6. Select the new directory and choose **OK**.

   The tab sheet shows the new directory as the destination for the selected file or directory.

## Copying Documents

Use the *SW* Document Manager to copy a document and all its associated files to a new location. You can't rename a document with the **Copy** operation. If you need to specify new names for files, use the **Rename** operation.

▶ **To copy a document**

1. Start the *SW* Document Manager.

2. In the **File Selection** box, select the document you want to copy.

   The document file must have a `.tex` extension.

3. Choose **Copy**.

4. In the **Copy To** dialog box, select the drive and the directory to which you want to copy the document.

   If you're working on a network, you may need to connect to a different network directory.

5. If you want to create a new directory to contain the copied document, choose **Create**, specify the directory name, and then choose **OK**.

   The program will create the new directory as a subdirectory in the currently selected directory.

6. In the **Copy** dialog box, select the associated document and graphics files you want to include in the operation.

7. If you want to change the destination of any associated files, choose **Change Destination**, specify a different destination for the selected files, and then choose **OK**.

8. In the **Copy** dialog box, choose **OK** to begin the copy operation.

9. If the destination directory contains a file of the same name, the program asks if you want to overwrite it.

   - Choose **Yes** to overwrite the file.
   - Choose **Yes to all** to overwrite all files with the duplicate names.
   - Choose **Skip** to exclude the file from the operation.
   - Choose **Cancel** to return to the **Copy** dialog box.

   A message box indicates when the copy operation is complete.

10. Choose **Close** to leave the *SW* Document Manager.

## Deleting Documents

When you no longer need a document, you should delete it to save disk space. Because *SW* documents can contain subdocuments and references to graphics files, you should use the *SW* Document Manager to delete a document and all associated files.

▶ **To delete a document and associated files**

1. Start the *SW* Document Manager.

2. In the **File Selection** box, select the document you want to delete.

3. Choose **Delete**.

   The *SW* Document Manager opens the **Delete** dialog box and selects for deletion all associated files with the following extensions: `.aaa-.aaz`, `.aut`, `.aux`, `.bak`, `.bbl`, `.bdx`, `.blg`, `.cdx`, `.dvi`, `.idx`, `.ilg`, `.ind`, `.lis`, `.lof`, `.log`, `.lot`, and `.toc`.

4. From the **Document** and **Graphics** tab sheets, exclude any files that you don't want the Document Manager to delete.

5. Choose **OK** to delete the document and all associated files.

   A message box indicates when the delete operation is complete.

6. Choose **Close** to leave the *SW* Document Manager.

## Renaming Files

Use the **Rename** operation to change the name of a file. Use the **Copy** operation to rename directories. When you rename a file, the *SW* Document Manager will correct all references to that file.

▶ **To rename a file**

1. Start the *SW* Document Manager.

2. In the **File Selection** box, select the document containing the subdocument or graphics file that you want to rename.

   The document file must have a `.tex` extension.

3. Choose **Rename**.

   The program opens the **Rename** dialog box:

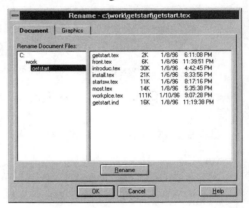

4. Select the associated document or graphics file you want to rename.

5. Choose **Rename**.

   The program opens the **Rename File** dialog box.

6. Specify a different file name for the selected file.

7. Choose **OK**.

   In the file section of the **Rename** dialog box, the original name is shown in parentheses next to the renamed file, like this:

   readthis(getstart.tex) 2K     1/8/96   6:11:08 PM

8. Choose **OK**.

   A message box indicates when the rename operation is complete.

9. Choose **Close** to leave the *SW* Document Manager.

## Cleaning up Generated Files

From time to time you should delete generated files to save disk space. Generated files have the same name as the document file but have different extensions.

▶ **To clean up generated files**

1. Start the *SW* Document Manager.

2. In the **File Selection** box, select the document for which you want to clean up the generated files.

3. Choose **Cleanup**.

   The *SW* Document Manager opens the **Cleanup** dialog box and automatically selects for deletion all associated files with the following extensions: `.aux`, `.blg`, `.dvi`, `.idx`, `.ilg`, `.lis`, `.lof`, `.log`, `.lot`, and `.toc`, and all but the latest file with an extension `.aaa-.aaz`.

4. Exclude any files that you don't want to be deleted in the cleanup operation.

5. Choose **OK** .

   A message box indicates when the cleanup operation is complete.

6. Choose **Close** to leave the *SW* Document Manager.

## Sending and Receiving Documents

When you send a document to another location on disk or by electronic mail, you must be sure to include not only the document itself but also any associated files and subdocuments. Use the *SW* Document Manager **Wrap** operation to gather in a single text file all the files that accompany the primary document file. Use the **Unwrap** operation to break the text file into separate files again. Documents that have been wrapped can be unwrapped on an *SW* system or, if *SW* is not available, using a text editor on a system that runs LaTeX.

▶ **To wrap a document**

1. Start the *SW* Document Manager.

2. In the **File Selection** box, select the document you want to wrap.

3. Choose **Wrap**.

   The *SW* Document Manager opens the **Wrap** dialog box, displays a default name for the wrapped document, and selects all associated document, graphics, and style files to be wrapped with your document.

4. If you want to change the name of the wrapped file, enter the name you want.

Remember that the wrapped file must have an .msg file extension.

5. Exclude any files you don't want to wrap with your document.

Remember that snapshots of Maple plots can be very large. If you are sending the document to someone who has *Scientific WorkPlace*, you do not need to include snapshots of Maple plots in the operation. You may also want to exclude standard *SW* styles.

6. Choose **OK**.

The program creates a text file that contains your document, all selected associated files, and instructions for using a text editor to re-create the original files in case the recipient of your document does not have *SW*. The Status bar indicates the progress of the wrap operation.

7. When the operation has finished, choose **Close** to leave the *SW* Document Manager.

▶ **To unwrap a document with the SW Document Manager**

1. Start the *SW* Document Manager.

2. From the **File Selection** box, select the document you want to unwrap.

The file you choose must have an .msg file extension.

3. Choose **Unwrap**.

The program opens the **Unwrap To** dialog box.

4. Select the destination directory for the unwrapped document.

5. Choose **OK**.

The program opens the **Unwrap** dialog box and indicates the destination directories for the document, graphics, and style files associated with the wrapped document.

6. Make any changes you want to the destination directories.

7. Exclude any files you don't want to unwrap with the document.

8. Choose **OK**.

9. If the destination directory contains a file of the same name as a file in the wrapped document, the program asks if you want to overwrite it.

   - Choose **Yes** to overwrite the file.
   - Choose **Yes to all** to overwrite all files with the duplicate names.
   - Choose **Skip** to exclude the file from the operation.
   - Choose **Cancel** to return to the **Copy** dialog box.

*SW* unwraps the document, placing the `.tex` file and its associated files in the specified destination directories.

10. When the operation has finished, choose **Close** to leave the *SW* Document Manager.

### ▶ To unwrap a document with a text editor

1. Open the `.msg` file with the text editor.

2. Follow the instructions contained in the file header.

### ▶ To send a wrapped document on disk

1. Wrap the document.

2. Exit *SW*.

3. Open the Windows File Manager.

4. Open the directory containing the wrapped document.

5. Check that the size of the wrapped document does not exceed the capacity of your disk.

6. Use the commands in the Windows File Manager to copy the wrapped document to your disk.

---

**Important**    Wrapping a document that contains graphics can create an `.msg` file that exceeds the capacity of a floppy disk. You may need to use a compression utility that permits breaking a file across more than one disk.

---

# 13 Customizing SW

*SW* is easy to use because of its straightforward approach to entering text and mathematics. You can make *SW* even more convenient to use by customizing the window, the tools, and the program defaults to suit the way you work.

## Customizing the Window

The size of the *SW* main window is variable. You can choose which toolbars and symbol panels appear in the window and where in the window they appear. You can have several documents open at the same time (see Chapter 2 "Opening and Closing Documents"), and you can arrange them to your liking in the main *SW* window. You can also maximize a document window or minimize it so that it appears as an icon in the main window. And you can customize the size and characteristics of the display in each document window as necessary.

### Sizing the SW Main Window

As in any Windows application, you can size the *SW* main window to your liking.

▶ **To enlarge or reduce the SW main window**

- On the Title bar, use the Maximize and Minimize buttons ▾ ▴ .
  −or−
- Double-click the Title bar.
  −or−
- From the Control box, choose **Maximize** or **Minimize**.
  −or−
- Use the mouse to drag the edges of the window to the size you want.
  When the *SW* main window is maximized, the sizing buttons on the Title bar appear
  like this: ▾ ▾ .

### Displaying the Toolbars

To customize your workplace, you can display or hide any of the *SW* toolbars. You can move them around on the main window, *docking* them in a permanent position or letting them *float* anywhere on the screen, even outside the *SW* main window. Floating toolbars have a title bar. You can also change the size of the buttons on the toolbars and turn off and on the display of the tooltips that describe each button's function. The program retains the way you arrange the toolbars from session to session.

▶ **To display or hide the toolbars**

1. From the View menu, choose Toolbars (ALT+V, T).

2. In the Toolbars dialog box, check the box next to the name of each toolbar you want to display.

3. Choose OK.

▶ **To dock a toolbar in the SW window**

1. Place the mouse pointer anywhere in the gray area surrounding the buttons on the toolbar.

2. Drag the toolbar to a new location in the gray area at the top or bottom or on the sides of the main window.

   Toolbars containing drop-down lists can't be docked on the sides of the main window.

▶ **To float a toolbar on the screen**

1. Place the mouse pointer anywhere in the gray area surrounding the buttons on the toolbar.

2. Drag the toolbar to a new location in the entry area of the main window or outside the main window.

▶ **To close a floating toolbar**

• Click the Close button in the upper-left corner of the toolbar.

▶ **To return the toolbars to their default positions and display**

1. From the View menu, choose Toolbars (ALT+V, T).

2. Choose Reset.

3. Choose OK or Close.

▶ **To display large buttons on the toolbars**

1. From the View menu, choose Toolbars (ALT+V, T).

2. Check Large Buttons.

3. Choose OK.

▶ **To toggle the display of tooltips on or off**

1. From the View menu, choose Toolbars (ALT+V, T).

2. Check or uncheck Show ToolTips.

3. Choose OK.

## Displaying the Symbol Panels

For faster access to symbols and characters, you can leave open the panels of symbols and characters available from the Symbol toolbar. You can dock the panels on the window or leave them floating anywhere you want. As with toolbars, *SW* retains the way you arrange the panels from session to session.

▶ **To display a symbol panel**

• On the Symbol toolbar, click the button for the panel you want.

▶ **To close a symbol panel**

• Click the Close box in the upper-left corner of the panel.
  –or–
• On the Symbol toolbar, click the button for the open panel.

▶ **To dock an open panel in the SW window**

1. Place the mouse pointer anywhere in the gray area surrounding the symbols on the panel.

2. Drag the toolbar to a new location in the gray areas at the top or bottom or on the sides of the main window.

▶ **To float an open panel on the screen**

1. Place the mouse pointer on the title bar of the panel.

2. Drag the toolbar to a new location in the entry area of the main window or outside the main window.

## Arranging Multiple Documents

If you have several documents open, you can arrange them in the window any way you want. If you maximize one document window, the other open windows remain open behind it.

### ▶ To maximize a document window

- Click the Maximize button in the upper-right corner of the document window.
  –or–
- Double-click the title bar of the document window.
  –or–
- Press CTRL+-, X.

  The program maximizes the document window. The Control box and the Restore and Minimize buttons for the maximized window appear on the Menu bar.

### ▶ To restore a maximized window to its original size

- On the right end of the Menu bar, click the Restore button  .
  –or–
- Press CTRL+-, R.

### ▶ To minimize a document window

- Click the Minimize button in the upper-right corner of the document window.
  –or–
- Press CTRL+-, N.

  In the main window, *SW* creates an icon containing the name of the document. If another document window is maximized, the minimized document is hidden.

### ▶ To restore the display of a minimized document window

- Double-click the icon for the document window.
  –or–
- Click the icon and then choose **Restore**.
  –or–
- From the **Window** menu, choose the window title of the minimized document.

### ▶ To arrange the open windows

- From the **Window** menu, choose
    - **Cascade** (ALT+W, C).
    - **Tile Horizontally** (ALT+W, H).
    - **Tile Vertically** (ALT+W, T).
  –or–
- With the mouse, drag the title bar of an open window to position it conveniently.

## Changing the Magnification in a Document Window

You can reduce or magnify the size of the document in each window from 50% normal size to 400% normal size. Normal size is an arbitrarily chosen size most people will use for ordinary work. Increased magnification is useful if you have difficulty discerning text at normal size or if you want to do detailed work in a complex mathematical expression. Decreased magnification can be helpful when dealing with very long mathematical expressions. Regardless of the size of the text, *SW* always breaks lines to fit the screen window, so there is no need to scroll horizontally. The one exception to this is when an unbreakable object is too large for the current window at the current view size.

You can display your document at normal size or at twice normal size. You can also customize the *working view*—the percentage of magnification you want the program to apply routinely when you view documents on the screen. The program applies the specified percentage to the normal display size. When you open a document, it is displayed using the magnification you used most recently. If you have several document windows open, you can use a different display size for each window.

▶ **To customize the working view size**

1. From the Tools menu, choose User Setup (ALT+L, U).

2. Choose the General tab.

3. In the Working View box, select the working view percentage you want.

4. Choose OK to set the percentage and record it in the View menu.

▶ **To change the display size of the active document window**

- From the View menu, choose Working (ALT+V, W), 100% (ALT+V, 1), or 200% (ALT+V, 2).
    - The 100% view size is the normal size.
    - The 200% view size is twice the normal size.
    - The working view size is nnn% of the normal size.
    - —or—

    a. From the Standard toolbar, choose the Zoom button 100% ▼ .
    b. Set the percentage of magnification you want for the active document window.

## Changing the Characteristics of the View

*SW* provides nonprinting characters, or *invisibles*, and guidelines for certain mathematical objects that facilitate the entry of text and mathematics. We suggest you turn the nonprinting elements on when you're creating or editing a document and off when you're more interested in reading it than making changes. You can display or hide nonprinting elements with the commands on the **View** menu. Some commands affect the display of all open documents; others, only the document in the active window. When you open a new document, it is displayed using the view settings you used most recently.

In *SW,* these elements can be displayed or hidden:

- Invisibles, or nonprinting characters, in text.
  Invisibles indicate the spaces between words and the ends of paragraphs, like this:

  This␣shows␣how␣paragraphs␣and␣spaces␣appear␣when␣invisibles␣are␣displayed.¶

- Lines that indicate the input boxes in mathematical structures.
  Input boxes indicate an empty place where you can type characters, symbols, or other mathematical objects. For example, the numerator and denominator boxes of a fraction before you enter text in either of these fields appears like this:

- Lines that indicate rows and columns in tables and matrices and the borders of displayed mathematics and multiline equations.
  *SW* normally shows a grid of the cells in tables and matrices to assist you in entering and editing content. The grid looks like this:

  Mathematical displays and multiline equations are boxed with nonprinting lines like this:

  $$x = a + b$$
  $$\leq c + d$$

- Field boxes for markers and index entries.
  Field boxes appear like this:

  index:  View Menu commands

### ▶ To display or hide nonprinting elements

- On the **View** menu, check the element you want to display
    - **Invisibles** (ALT+V, I).
    - **Matrix Lines** (ALT+V, M).
    - **Input Boxes** (ALT+V, B).
    - **Index Fields** (ALT+V, X).
    - **Marker Fields** (ALT+V, K).

▶ **To display or hide nonprinting characters in text**

- On the Standard toolbar, click the Show/Hide Nonprinting button .

## Customizing the Tools and Defaults

You can also customize certain *SW* tools, including the function key assignments and mouse keypresses, and you can set program defaults for options related to the document, text, and mathematics interface; document style; graphics; file storage; and computations.

## Changing the Assignments of Tags to Function Keys

You can assign item, section, and text tags to the function keys on your keyboard so you can apply the tags more quickly. You can save tag assignments along with style information so they are loaded as part of the style. If you change the key assignments for a style, the new assignments are in effect for all other documents of the same style. You can also save tag assignments globally. If you make global assignments, they are in effect for all documents. When a function key has been assigned both globally and with style information, the assignment in the style information takes precedence.

▶ **To change a function key assignment**

1. From the Tag menu, choose Function Keys (ALT+T, F).

   The program opens the Tag Key Assignments dialog box:

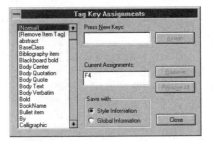

2. From the list of tags, select the tag you want to assign to a function key.

3. In the Save with box, choose the environment for the settings.

4. Position the insertion point in the Press New Keys box.

5. Press the function key you want.

   You can use *modifiers* such as CTRL, ALT, and SHIFT. The current assignment of the key is displayed in the dialog box.

6. Choose **Assign**.

If the key you choose is already assigned to a tag, the program clears the old assignment.

7. Choose **Close**.

▶ **To clear a tag assignment**

1. From the Tag menu, choose **Function Keys** (ALT+T, F).

2. From the list of tags, select the tag whose assignment you want to clear.

3. If you want to clear a single assignment for the tag, select it from the **Current Assignments** box and choose **Remove**.

4. If you want to clear all assignments for the tag, choose **Remove All**.

5. Choose **Close**.

## Customizing the Defaults

The tab sheets in the **User Setup** dialog box contain the controls for customizing the defaults related to the document, text, and mathematics interface; document style; graphics; file storage; and computations. You set most of the defaults by checking boxes and buttons on and off, by entering numbers to indicate settings, or by typing information.

▶ **To customize the** SW **defaults**

1. From the Tools menu, choose **User Setup** (ALT+L, U).

2. Choose the tab for the kind of default you want to set.

3. Make the settings you want.

4. Choose **OK**.

---

**Important**    Remember that when you choose **OK** in a tabbed dialog box, you accept the settings made on all tab sheets. When you choose **Cancel**, you discard the settings made on all tab sheets.

---

## Customizing General Defaults

Use the General tab sheet to change defaults relating to general properties of *SW*.

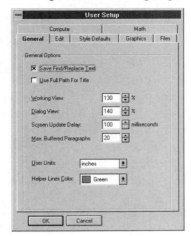

| Control | Use |
|---------|-----|
| Save Find/Replace Text | Retain the text of find and replace search patterns from session to session. |
| Use Full Path For Title | Display the full path of a document in the title bar. |
| Working View | Set the percentage of magnification for all documents you open. |
| Dialog View | Set the percentage of magnification used in dialog boxes. |
| Screen Update Delay | Set the frequency with which *SW* updates the screen. |
| Max Buffered Paragraphs | Set the number of paragraphs *SW* keeps internally. |
| User Units | Set the default unit of measure for *SW* measurements. |
| Helper Lines Color | Set the color used to display lines in nonprinting elements. |

## Customizing Edit Defaults

Use the Edit tab sheet to change defaults relating to general editing of *SW* documents.

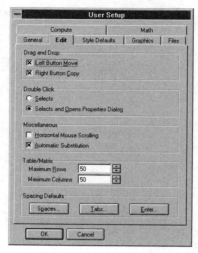

| Control | Use |
|---|---|
| **Drag and Drop** | |
| Left Button Move | Use the left mouse button to move a selection. |
| Right Button Copy | Use the right mouse button to copy a selection. |
| **Double Click** | |
| Selects | Double-click to select an object. |
| Selects and Opens Properties Dialog | Double-click to select an object and open the c sponding properties dialog box. |
| **Miscellaneous** | |
| Horizontal Mouse Scrolling | Controls whether or not *SW* scrolls the screen zontally when dragging the mouse left and right the edge of the screen. |
| Automatic Substitution | Enable automatic substitution. |
| **Table/Matrix** | |
| Maximum Rows | Set the maximum number of rows allowed in a tab matrix. |
| Maximum Columns | Set the maximum number of columns allowed in a t or matrix. |
| **Spacing Defaults** | |
| Spaces | Set the default for handling two spaces in successi |
| Tabs | Set the default for handling a TAB keypress in text |
| Enter | Set the default for handling an ENTER keypress at beginning of a paragraph. |

### Customizing Style Defaults

Starting *SW* automatically opens an empty start-up document. Use the **Style Default** tab sheet to set the default document type and style for the start-up document.

| Control | Use |
| --- | --- |
| Document Type | Set the default document type. |
| Styles | Set the default document style. |
| Style List Filter | Set the kinds of styles you want to have listed automatically when you open the **Predefined Styles** dialog box. |
| Version | Set the version of LaTeX for which you want to display available styles when you open the **Predefined styles** dialog box. |
| Preview Sample | Preview a sample of the default style you select. |

## Customizing Graphics Defaults

Use the Graphics tab sheet to set the defaults for working with graphics.

| Control | Use |
|---|---|
| **Global Settings** | |
| Max Graphics Memory Usage | Set an upper limit on the graphics memory used by *SW*. |
| Max Inactive Graphics Files | Set the maximum allowable number of inactive graphics files. |
| **Screen Attributes** | Set the default for the appearance of graphics on the screen: |
| Picture with Frame | Display both the picture and the surrounding frame. |
| Frame Only | Display only the frame. |
| Picture Only | Display only the picture. |
| Iconify | Minimize the picture so that it appears as an icon. |
| **Size** | |
| Width | Set the width of new graphics to this dimension. |
| Height | Set the height of new graphics to this dimension. |
| Units | Set the default unit of measure for new graphics. |
| **Placement** | |
| In Line | Place all new graphics in line. |
| Displayed | Center all new graphics on a separate line. |
| Floating | Allow all new graphics to float. |
| **Miscellaneous** | |
| Maintain Aspect Ratio | When sizing a graphic, maintain its proportions. |
| Fit to Default Size | Fit all graphics to the default size. |
| Print with Frame | Print all graphics with a frame. |
| Print with Picture | Print all graphics with a picture. |
| **Floating Placement** | |
| Here | Float graphics where the frame resides. |
| On a page of Floats | Float graphics on a separate page of floating objects. |
| Top of Page | Float graphics at the top of a page. |
| Bottom of Page | Float graphics at the bottom of a page. |

## Customizing File Defaults

Use the Files tab sheet to set the defaults for working with files and directories.

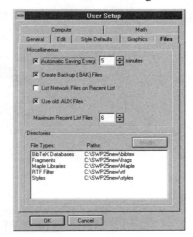

| Control | Use |
|---------|-----|
| Automatic Saving Every | Set the frequency of automatic saving. |
| Create Backup (.BAK) Files | Create backup files each time the document is saved. |
| List Network Files on Recent List | Include on the list of recently opened documents any documents that reside on network drives. |
| Use old .AUX Files | Save the auxiliary files created when documents are compiled. |
| Maximum Recent List Files | Set the maximum number of documents that can be included on the list of recently opened files. |
| Directories | The drive and directory locations of file types used by *SW*. |
| Modify | Modify drive and directory locations. |

### Customizing Math Defaults

Use the **Math** tab sheet to set the defaults for creating mathematical objects.

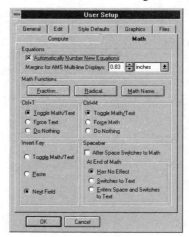

| Control | Use |
|---|---|
| **Equations** | |
| Automatically Number New Equations | Automatically create an equation number for each new equation line. |
| Margins for AMS Multi-line Displays | Set the distance from the left and right screen edges of a single equation displayed on multiple lines. |
| **Math Functions** | |
| Fraction | Set the default size and line type for fractions. |
| Radical | Set the default for new radicals to be with or without roots. |
| Math Name | Set the defaults for placing limits in functions and for adding newly defined math names to the list of recognized math names. |
| **Ctrl+T** and **Ctrl+M** | |
| Toggle Math/Text | Toggle between math and text. |
| Force Text or Force Math | Change to text or to math. |
| Do Nothing | |
| **Insert Key** | |
| Toggle Math/Text | Toggle between math and text. |
| Paste | Paste the current selection. |
| Next Field | Move to the next field. |
| **Spacebar** | |
| After Space Switches to Math | Set the second of two SPACEBAR keypresses to toggle to math. |
| **At End of Math** | |
| Has No Effect | Set a SPACEBAR keypress at the end of a string of mathematics so that it has no effect. |
| Switches to Text | Switch to text when SPACEBAR keypress occurs at end of a string of mathematics. |
| Enters Space and Switches to Text | Enter space and switch to text after SPACEBER keypress occurs at end of a string of mathematics. |

### Customizing Computation Defaults

Use the **Compute** tab sheet to set the defaults for mathematical computations.

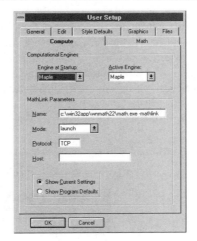

| Control | Use |
|---|---|
| **Computational Engines** | |
| Engine at Startup | Choose the computational software in effect when you open *SW*. |
| Active Engine | Choose the computational software for current use. |
| **MathLink Parameters** | |
| Name | Set the full path to the `math.exe` file in your Mathematica directory. |
| Mode | Set the parent-child relationship of the communication link to Mathematica. |
| Protocol | Set the data-transfer protocol used in the Mathematica communication. |
| Host | Set the machine on which the other link partner resides. |
| Show Current Settings | Show the Mathlink parameters read from the `.ini` file. |
| Show Program Defaults | Show the Mathlink parameters hard-coded in *SW*. |

## Customizing Automatic Substitution

We use the term *automatic substitution* to describe the feature that allows you to substitute expressions for sequences of letters and numbers in mathematics. Some substitutions are predefined. For example, if you type **cos** when the insertion point is in mathematics, the *co* appears on the screen in italics until you type the *s*, and then the entire function name *cos* changes to an upright cos and is grayed. The function cos has been substituted for the three-letter sequence *c*, *o*, and *s*.

You can define and edit your own substitution sequences and expressions. For example, you might want to substitute an expression such as $\sum_{n=1}^{10}$ for the sequence **sum10**.

If you type **sum10** when the insertion point is in mathematics, the $sum$ appears in italics and the 1 in red until you type the 0. Then, the entire sequence is replaced with $\sum_{n=1}^{10}$.

You can enable or disable the automatic substitution of sequences at any time. If you disable automatic substitution, the program does not substitute expressions for special sequences of letters or numbers. If you disable automatic substitution and type **cos** when the insertion point is in mathematics, the letters are not displayed in gray, upright type nor are they substituted by the function. Similarly, if you type your customized sequence **sum10**, the program does not substitute the expression $\sum_{n=1}^{10}$.

▶ **To define a new automatic substitution sequence**

1. From the Tools menu, choose Automatic Substitution (ALT+L, T).

   The program opens the Automatic Substitution dialog box:

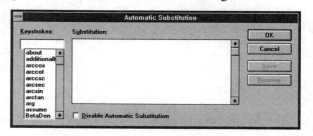

2. In the Keystrokes box, type the substitution sequence.

   The sequence must consist of alphabetic and numeric characters only. When you type the character sequence in mathematics, the program automatically replaces it with the expression you enter in the Substitution box.

3. In the Substitution box, type the replacement expression.

   The expression can be any valid mathematics expression. You can use the menu commands and toolbar buttons to enter parts of the expression, or you can paste a mathematical expression from the clipboard to the Substitution box.

4. Choose Save.

5. Enter and save any additional substitution sequences and replacement expressions you want.

6. Choose OK.

▶ **To edit a replacement expression**

1. From the Tools menu, choose Automatic Substitution (ALT+L, T).

2. From the Automatic Substitution dialog box, select the substitution sequence for the expression you want to edit.

3. Edit the expression in the Substitution box.

   You can use the menu commands and the toolbar buttons to edit the expression.

4. Choose Save.

5. Edit and save any other substitution sequences and replacement expressions you want.

6. Choose OK.

▶ **To delete an automatic substitution sequence and replacement expression**

1. From the Tools menu, choose Automatic Substitution (ALT+L, T).

2. From Automatic Substitution dialog box, select the substitution sequence for the expression you want to delete.

3. Choose Remove to delete the substitution sequence and the replacement expression.

4. Choose OK.

▶ **To turn automatic substitution on or off**

• Check the Disable Automatic Substitution box in the Automatic Substitution dialog box.
  –or–
• In the Edit tab sheet of the User Setup dialog box, check the Automatic Substitution box.

# A   Toolbar Buttons and Menu Commands

## Standard Toolbar

| Button | Menu / Command | Use |
|---|---|---|
| | File / New | Open a new document |
| | File / Open | Open an existing document |
| | File / Save | Save the active document |
| | File / Print | Print the active document |
| | File / Preview | Preview the active document |
| | Tools / Spelling | Check the spelling of the active document |
| | Edit / Cut | Move a selection to the clipboard |
| | Edit / Copy | Copy a selection to the clipboard |
| | Edit / Paste | Paste the information on the clipboard into the document |
| | Edit / Undo Deletion | Undo the most recent deletion |
| | Insert / Table | Insert a table |
| | Edit / Properties | Examine the properties of the selected object |
| | View / Invisibles | Show or hide nonprinting characters |
| 100% | | Change the size of a document on the screen |

## Math Toolbar

| Button | Menu / Command | Use |
|--------|----------------|-----|
| ⊟ | Insert / Fraction | Insert a fraction |
| √□ | Insert / Radical | Insert a radical |
| $N^x$ | Insert / Superscript | Insert a superscript |
| $N_x$ | Insert / Subscript | Insert a subscript |
| (□) | Insert / Brackets | Insert expanding parentheses |
| [□] | Insert / Brackets | Insert expanding brackets |
| Σ | Insert / Operator | Insert a summation operator |
| ∫ | Insert / Operator | Insert an integral operator |
| ≡ | Insert / Display | Insert a mathematical display |
| T | Insert / Math/Text | T to enter text; M to enter mathematics |
| Σ∫ | Insert / Operator | Insert mathematical operators such as summations, integrals, and products |
| ()[] | Insert / Brackets | Insert expanding brackets and enclosures in matched or unmatched pairs |
| ⠿ | Insert / Matrix | Insert a matrix or vector |
| sin cos | Insert / Math Name | Insert a mathematical function (e.g., sin, lim) |
| e..m | Insert / Spacing | Insert thick, thin, required, or other horizontal spacing between characters or symbols |
| (:) | Insert / Binomial | Insert binomials or generalized fractions |
| ⊟ | Insert / Label | Insert labels for expressions or formulas |
| ↔ | Insert / Decoration | Insert bars, arrows, or braces over or under expressions, or insert boxes around expressions |

## Symbol Toolbar

| Button | Use |
|--------|-----|
| αβ | Lowercase Greek characters |
| ΦΨ | Uppercase Greek characters |
| ± ÷ | Binary operations |
| ≤ ⊆ | Binary relations |
| ≠ ∉ | Negated symbols |
| ↔ ↑ | Arrows |
| ∞ ∂ | Miscellaneous symbols |
| ) [ | Special delimiters |
| £ ¹ | Latin1 Unicode characters |
| Ł ĕ | Latin Extended-A Unicode characters |
| " " | General Punctuation Unicode characters |

## Compute Toolbar

| Button | Menu / Command |
|--------|----------------|
| =? | Maple / Evaluate |
| #? | Maple / Evaluate Numerically |
| x+x=2x | Maple / Simplify |
| ● | Maple / Expand |
| ╪ | Maple / Plot 2D |
| ╪ | Maple / Plot 3D |
| ƒ(x) | Maple / Define / New Definition |
| ƒ | Maple / Define / Show Definitions |
| ▦ | Maple / Matrix / Fill Matrix |

## Navigate Toolbar

| Button | Menu / Command | Use |
|--------|----------------|-----|
| ◀ | Go / Previous Section | Go to the previous section |
| ▷ | Go / Next Section | Go to the next section |
| ↰ | Go / History Back | Go to the source of the last jump |
| ✦ | Go / Goto Marker | Go to the paragraph containing the marker |

## Field Toolbar

| Button | Menu / Command | Use |
|--------|----------------|-----|
| " | Insert / Field / Citation | Create a citation for an item listed in the bibliography |
| ≣ | Insert / Field / Note | Create a footnote or marginal note |
| ≣ | Insert / Field / Index Entry | Create an index entry |
| ✦ | Insert / Field / Marker | Create a marker for a document page or part |
| ≣ | Insert / Field / Cross Reference | Create a reference to a document page number or part |
| ≣ | Insert / Field / Hypertext Link | Create a hypertext link to any marker or any object with a key |
| RUN | Insert / Field / External Program Call | Create a link to another program from inside the $SW$ document |
| TeX | Insert / Field / TeX | Insert a field for TeX commands |
| ⊟ | Insert / Field / Subdocument | Create a subdocument within a master document |
| ▤ | Insert / Field / Bibliography | Insert a BibTeX bibliography |

# B  Keyboard Shortcuts and Key Prefixes

## Keyboard Shortcuts

| To move | | Press |
|---|---|---|
| To the left | | LEFT ARROW |
| To the right | | RIGHT ARROW (or SPACEBAR in math) |
| To the start of the next word | | CTRL+RIGHT ARROW |
| To the start of the previous word | | CTRL+LEFT ARROW |
| Up | | UP ARROW |
| Down | | DOWN ARROW |
| To the start of the line | | HOME |
| To the end of the line | | END |
| To the next field inside a template | | TAB |
| | or | ARROW KEYS |
| To the previous field inside a template | | SHIFT+TAB |
| | or | ARROW KEYS |
| To the outside of a template | | RIGHT ARROW |
| | or | LEFT ARROW |
| | or | SPACEBAR |
| Next screen | | PAGE DOWN |
| Previous screen | | PAGE UP |
| To the beginning of the document | | CTRL+HOME |
| To the end of the document | | CTRL+END |
| To the source of the most recent jump | | CTRL+ALT+LEFT ARROW |
| To the next paragraph | | CTRL+ALT+N |
| To the previous paragraph | | CTRL+ALT+P |

| To | Press |
|---|---|
| Toggle math/text | CTRL+M |
| | or    CTRL+T |
| Negate the character to the left | CTRL+N |
| Undo the last deletion | CTRL+Z |
| Cut a selection to the clipboard | CTRL+X |
| Copy a selection to the clipboard | CTRL+C |
| Paste from the clipboard | CTRL+V |
| Enter a line break | SHIFT+ENTER |
| Refresh the screen | ESC |

| To select | Press |
|---|---|
| The object or symbol to the left of the insertion point | SHIFT+LEFT ARROW |
| The object or symbol to the right of the insertion point | SHIFT+RIGHT ARROW |
| The word to the right of the insertion point | CTRL+SHIFT+RIGHT ARROW |
| The word to the left of the insertion point | CTRL+SHIFT+LEFT ARROW |
| The smallest object that contains both the point at which the insertion point starts and the point to which the insertion point moves when you press UP ARROW | SHIFT+UP ARROW |
| The smallest object that contains both the point at which the insertion point starts and the point to which the insertion point moves when you press DOWN ARROW | SHIFT+DOWN ARROW |
| Everything between the insertion point and the start of the line | SHIFT+HOME |
| Everything between the insertion point and the end of the line | SHIFT+END |
| The following screen | SHIFT+PAGE DOWN |
| The previous screen | SHIFT+PAGE UP |
| Everything between the insertion point and the start of the document | CTRL+SHIFT+HOME |
| Everything between the insertion point and the end of the document | CTRL+SHIFT+END |

| To enter mathematical objects | | Press |
|---|---|---|
| Fraction | | CTRL+F |
| | or | CTRL+/ |
| | or | CTRL+1 |
| Radical | | CTRL+R |
| | or | CTRL+2 |
| Superscript | | CTRL+H |
| | or | CTRL+UP ARROW |
| | or | CTRL+3 |
| Subscript | | CTRL+L |
| | or | CTRL+DOWN ARROW |
| | or | CTRL+4 |
| Summation | | CTRL+7 |
| Integral | | CTRL+I |
| | or | CTRL+8 |
| Brackets | | CTRL+9 |
| | or | CTRL+0 |
| | or | CTRL+( |
| | or | CTRL+) |
| | or | CTRL+5 |
| Square brackets | | CTRL+9 |
| | or | CTRL+] |
| | or | CTRL+6 |
| Braces | | CTRL+{ |
| | or | CTRL+} |
| Display | | CTRL+D |
| 2 x 2 Matrix | | CTRL+A |
| Product | | CTRL+P |
| Absolute value | | CTRL+| |
| Thin space | | CTRL+SPACEBAR |
| Thick space | | SHIFT+SPACEBAR |
| Em space | | CTRL+SHIFT+SPACEBAR |
| " (double open quote) | | Single open quote (') twice |
| - (intraword dash or hyphen) | | Hyphen (-) |
| – (en dash) | | Hyphen (-) twice |
| — (em dash) | | Hyphen (-) three times |
| ¿ (inverted question mark) | | ? followed by ' |
| ¡ (inverted exclamation point) | | ! followed by ' |

| To enter accents | Press |
|:---:|:---|
| ^ | CTRL+^ (CTRL+SHIFT+6) |
| ~ | CTRL+~ (CTRL+SHIFT+`) |
| ´ | CTRL+' |
| ` | CTRL+` |
| . | CTRL+. |
| .. | CTRL+" (CTRL+SHIFT+') |
| -- | CTRL+= |
| → | CTRL+− |

## Key Prefixes

| To enter | Press ctrl+s then press | To enter | Press ctrl+s then press |
|:---:|:---:|:---:|:---:|
| → | 1 | ⊂ | c |
| ↑ | 2 | ∨ | v |
| ← | 3 | • | b |
| ↓ | 4 | ∇ | n |
| ⊇ | 5 | ⇒ | ! |
| ∩ | 6 | ⇑ | @ |
| ⊆ | 7 | ⇐ | # |
| ∪ | 8 | ⇓ | $ |
| (□) | 9 or 0 or ( or ) | ⊃ | % |
| ≡ | - | ≅ | _ |
| ≠ | = | ± | + |
| ≈ | w | ℵ | W |
| ∈ | e | ∉ | E |
| √□ | r or R | ∞ | I |
| ⊗ | t or T | ℘ | P |
| ∫ | i | {□} | { or } |
| ∅ | o | ∀ | A |
| ∏ | p | ⊕ | S |
| [□] | [ or ] | ◇ | D |
| ∠ | a | ÷ | X |
| ∑ | s | · | C |
| ∂ | d | ∧ | V |
| □/□ | f or F | ¬ | N |
| □ | h or H | ≤ | < |
| □ | l or L | ≥ | > |
| × | x | ∃ | z |
| □□/□□ | m | | |

| To enter | | Press ctrl+g then press | To enter | | Press ctrl+g then press |
|---|---|---|---|---|---|
| alpha | $\alpha$ | a | xi | $\xi$ | x |
| beta | $\beta$ | b | | $\Xi$ | X |
| gamma | $\gamma$ | g | pi | $\pi$ | p |
| | $\Gamma$ | G | | $\Pi$ | P |
| delta | $\delta$ | d | rho | $\rho$ | r |
| | $\Delta$ | D | sigma | $\sigma$ | s |
| epsilon | $\varepsilon$ | e | | $\Sigma$ | S |
| zeta | $\zeta$ | z | tau | $\tau$ | t |
| eta | $\eta$ | h | upsilon | $\upsilon$ | u |
| theta | $\theta$ | y | | $\Upsilon$ | U |
| | $\Theta$ | Y | phi | $\phi$ | f |
| iota | $\iota$ | i | | $\Phi$ | F |
| kappa | $\kappa$ | k | chi | $\chi$ | q |
| lambda | $\lambda$ | l | psi | $\psi$ | c |
| | $\Lambda$ | L | | $\Psi$ | C |
| mu | $\mu$ | m | omega | $\omega$ | w |
| nu | $\nu$ | n | | $\Omega$ | W |

# ANSI Characters

## ANSI Characters

| | | | | | | | | | | | |
|---|---|---|---|---|---|---|---|---|---|---|---|
| 160 | space | 176 | ° | 192 | À | 208 | Ð | 224 | à | 240 | ð |
| 161 | ¡ | 177 | ± | 193 | Á | 209 | Ñ | 225 | á | 241 | ñ |
| 162 | ¢ | 178 | $^2$ | 194 | Â | 210 | Ò | 226 | â | 242 | ò |
| 163 | £ | 179 | $^3$ | 195 | Ã | 211 | Ó | 227 | ã | 243 | ó |
| 164 | ¤ | 180 | ´ | 196 | Ä | 212 | Ô | 228 | ä | 244 | ô |
| 165 | ¥ | 181 | $\mu$ | 197 | Å | 213 | Õ | 229 | å | 245 | õ |
| 166 | ¦ | 182 | ¶ | 198 | Æ | 214 | Ö | 230 | æ | 246 | ö |
| 167 | § | 183 | · | 199 | Ç | 215 | × | 231 | ç | 247 | ÷ |
| 168 | ¨ | 184 | ¸ | 200 | È | 216 | Ø | 232 | è | 248 | ø |
| 169 | © | 185 | $^1$ | 201 | É | 217 | Ù | 233 | é | 249 | ù |
| 170 | $^a$ | 186 | º | 202 | Ê | 218 | Ú | 234 | ê | 250 | ú |
| 171 | « | 187 | » | 203 | Ë | 219 | Û | 235 | ë | 251 | û |
| 172 | ¬ | 188 | ¼ | 204 | Ì | 220 | Ü | 236 | ì | 252 | ü |
| 173 | – | 189 | ½ | 205 | Í | 221 | Ý | 237 | í | 253 | ý |
| 174 | ® | 190 | ¾ | 206 | Î | 222 | þ | 238 | î | 254 | Þ |
| 175 | ¯ | 191 | ¿ | 207 | Ï | 223 | ß | 239 | ï | 255 | ÿ |

## TeX Commands

► **To enter TEXcommands**

1. Hold down the CTRL key.

2. Type the command name.

3. Release the CTRL key.

### Decorations

| | |
|---|---|
| fbox | underbrace |
| frame | underleftarrow |
| overbrace | underleftrightarrow |
| overleftarrow | underline |
| overleftrightarrow | underrightarrow |
| overline | widehat |
| overrightarrow | widetilde |

### Big Operators

| | | |
|---|---|---|
| bigcap | biguplus | iint |
| bigcup | bigvee | int |
| bigodot | bigwedge | oint |
| bigoplus | coprod | prod |
| bigotimes | idotsint | sum |
| bigskip | iiiint | |
| bigsqcup | iiint | |

### Spaces and Breaks

| | |
|---|---|
| allowbreak | nolinebreak |
| linebreak | pagebreak |
| mathstrut | qquad |
| medskip | quad |
| negthinspace | smallskip |
| newline | strut |
| newpage | thinspace |

### Other Objects

frac
matrix — creates 2x2 matrix

## Binary Operators

| | | | | |
|---|---|---|---|---|
| amalg | boxtimes | dag | lozenge | star |
| ast | bullet | dagger | ltimes | times |
| bigcirc | Cap | ddag | mp | triangledown |
| bigstar | cap | ddagger | odot | triangleleft |
| bigtriangledown | cdot | diamond | ominus | triangleright |
| bigtriangleup | cdotp | div | oplus | unlhd |
| blacklozenge | centerdot | divideontimes | oslash | unrhd |
| blacksquare | circ | dotminus | otimes | uplus |
| blacktriangle | circledast | dotplus | pm | vartriangleleft |
| blacktriangledown | circledcirc | dotsquare | rhd | vartriangleright |
| blacktriangleleft | circleddash | doublecap | rtimes | vee |
| blacktriangleright | Cup | doublecup | setminus | wedge |
| boxdot | cup | land | smallsetminus | wr |
| boxminus | curlyvee | lhd | sqcap | |
| boxplus | curlywedge | lor | sqcup | |

## Open/Close

| Open | Close |
|---|---|
| lceil | rceil |
| langle | rangle |
| lfloor | rfloor |
| ulcorner | urcorner |
| llcorner | lrcorner |

## Lowercase Greek

| | | |
|---|---|---|
| alpha | mu | upsilon |
| beta | nu | varepsilon |
| chi | omega | varkappa |
| delta | phi | varphi |
| epsilon | pi | varpi |
| eta | psi | varrho |
| gamma | rho | varsigma |
| iota | sigma | vartheta |
| kappa | tau | xi |
| lambda | theta | zeta |

## Uppercase Greek

| | |
|---|---|
| Delta | Psi |
| Gamma | Sigma |
| Lambda | Theta |
| Omega | Upsilon |
| Phi | Xi |
| Pi | |

## Symbols of Type Ord

| | | | | |
|---|---|---|---|---|
| aa | clubsuit | exists | oe | surd |
| AA | dh | forall | OE | th |
| ae | DH | hbar | partial | TH |
| AE | digamma | Im | Re | therefore |
| aleph | dj | infty | restriction | vartriangle |
| angle | DJ | maltese | smallint | wp |
| backepsilon | ell | mho | spadesuit | |
| Bbbk | emdash | nabla | ss | |
| because | emptyset | ng | straightepsilon | |
| bracevert | endash | NG | suchthat | |

## Additional Symbols of Type Ord

| | | | | |
|---|---|---|---|---|
| backprime | degree | guilsinglleft | prime | textquotedblright |
| backslash | diagdown | guilsinglright | quotedblbase | triangle |
| beth | diagup | heartsuit | quotesinglbase | varnothing |
| Box | Diamond | hslash | rbrace | vdots |
| cdots | diamondsuit | imath | rbrack | vert |
| cents | dots | lbrace | registered | Vert |
| checkmark | extexclamdown | lbrack | rq | yen |
| circledR | Finv | ldots | sharp | |
| circledS | flat | lnot | sphericalangle | |
| complement | Game | lq | square | |
| copyright | gimel | measuredangle | textcurrency | |
| daleth | guillemotleft | natural | textquestiondown | |
| ddots | guillemotright | pounds | textquotedblleft | |

# Binary Relations

| | | | | |
|---|---|---|---|---|
| approx | Gg | longmapsto | nvdash | subsetneq |
| approxeq | ggg | Lsh | nVdash | subsetneqq |
| asymp | gggtr | lvertneqq | nvDash | succ |
| backcong | gnapprox | mapsto | nVDash | succapprox |
| backsim | gneq | mid | owns | succcurlyeq |
| backsimeq | gneqq | mlcp | parallel | succeq |
| barwedge | gnsim | models | perp | succnapprox |
| between | gtrapprox | multimap | pitchfork | succneqq |
| bot | gtrdot | ncong | prec | succnsim |
| bowtie | gtreqless | ne | precapprox | succsim |
| bumpeq | gtreqqless | neq | preccurlyeq | Supset |
| Bumpeq | gtrless | nexists | preceq | supset |
| circeq | gtrsim | ngeq | precnapprox | supseteq |
| Colon | gvertneqq | ngeqq | precneqq | supseteqq |
| colon | iff | ngeqslant | precnsim | supsetneq |
| cong | in | ngtr | precsim | supsetneqq |
| curlyeqprec | intercal | ni | propto | thickapprox |
| curlyeqsuc | intprod | nleq | proves | thicksim |
| dashv | Join | nleqq | questeq | toea |
| ddotseq | le | nleqslant | Relbar | toooooo |
| doteq | leadsto | nless | relbar | top |
| Doteq | leftthreetimes | nmid | rightthreetimes | tosa |
| doteqdot | leq | notin | risingdotseq | trianglelefteq |
| doublebarwedge | leqq | nparallel | Rsh | triangleq |
| eqcirc | leqslant | nprec | shortmid | trianglerighteq |
| eqcolon | lessapprox | npreceq | shortparallel | varpropto |
| eqsim | lessdot | nshortmid | sim | varsubsetneq |
| eqslantgtr | lesseqgtr | nshortparallel | simeq | varsubsetneqq |
| eqslantless | lesseqqgtr | nsim | smallfrown | varsupsetneq |
| equiv | lessgtr | nsubseteq | smallsmile | varsupsetneqq |
| fallingdotseq | lesssim | nsubseteqq | smile | vdash |
| forces | ll | nsucc | sqsubset | Vdash |
| frown | Ll | nsucceq | sqsubseteq | vDash |
| ge | lll | nsupseteq | sqsupset | veebar |
| geq | lllless | nsupseteqq | sqsupseteq | Vvdash |
| geqq | lnapprox | ntriangleleft | subset | wedgeq |
| geqslant | lneq | ntrianglelefteq | Subset | |
| gets | lneqq | ntriangleright | subseteq | |
| gg | lnsim | ntrianglerighteq | subseteqq | |

## Arrows

arrowvert
Arrowvert
bkarow
circlearrowleft
circlearrowright
curvearrowleft
curvearrowright
dasharrow
dashleftarrow
dashrightarrow
dbkarow
Downarrow
downarrow
downdownarrows
downharpoonleft
downharpoonright
drbkarow
hksearow
hkswarow
hookleftarrow
hookrightarrow
leftarrow
Leftarrow
leftarrowtail
leftharpoondown
leftharpoonup

leftleftarrows
leftrightarrow
Leftrightarrow
leftrightarrows
leftrightharpoons
leftrightsquigarrow
Lleftarrow
longleftarrow
Longleftarrow
Longleftrightarrow
longleftrightarrow
Longrightarrow
longrightarrow
looparrowleft
looparrowright
nearrow
nleftarrow
nLeftarrow
nLeftrightarrow
nleftrightarrow
nrightarrow
nRightarrow
nwarrow
Rightarrow
rightarrow
rightarrowtail

rightharpoondown
rightharpoonup
rightleftarrows
rightleftharpoons
rightrightarrows
rightsquigarrow
Rrightarrow
searrow
swarrow
twoheadleftarrow
twoheadrightarrow
Uparrow
uparrow
Updownarrow
updownarrow
upharpoonleft
upharpoonright
upuparrows

# Index

238

*Are you using Scientific WorkPlace™ on a college campus?*

**If you currently teach, or plan to teach, Differential Equations, you owe it to yourself and your students to send for a review copy of:**

Preliminary Edition

# Differential Equations

**by Boston University's Paul Blanchard, Robert L. Devaney, and Glen Richard Hall**

There is a technological revolution sweeping through the mathematics curriculum, and indeed, there is no course for which this revolution is more significant than the differential equations course. Based on their experiences in the National Science Foundation-funded Boston University Differential Equations Project, these authors have written the first reform differential equations text with a dynamical systems flavor. In addition to its significant and consistent use of technology and emphasis on using programs such as Scientific WorkPlace™ to assist with problem solving, this innovative text:

• strikes a balance among analytic, numeric, and graphical techniques • emphasizes modeling • stresses qualitative theory throughout the course • presents linear and nonlinear systems in parallel • includes an introduction to discrete models of dynamical systems • and emphasizes interpretation, qualitative description, lab reports, and written explanations so users gain a greater conceptual understanding of the subject.

This fresh approach to teaching differential equations is so forward thinking that more than 30 schools adopted pre-publication versions of the text. We invite you to take advantage of this opportunity to receive a complimentary review copy of Blanchard, Devaney, and Hall's revolutionary **Differential Equations** by returning the reply form below.

------------------------------------------------------------------------

## REPLY ORDER FORM

_____ Yes, I would like to receive a complimentary review copy of Blanchard, Devaney, and Hall's **Differential Equations, Preliminary Edition**. (I am enclosing with this order form a copy of our department letterhead bearing my signature.)

PLEASE SHIP MY REVIEW COPY TO:

NAME:_____

INSTITUTION and DEPARTMENT:_____

STREET ADDRESS:_____

CITY:_____ STATE:_____ ZIP + 4: _____

MAIL TO: Brooks/Cole Publishing Company, Dept. Creating Documents Offer, 511 Forest Lodge Road, Pacific Grove, CA, 93950-5098

*Make maximum use of* **Scientific WorkPlace™** *as a computational tool with*

# Doing Mathematics With Scientific WorkPlace,
## Revised Edition

by Darel Hardy (Colorado State University) and Carol Walker (New Mexico State University)

Designed to show users how to get the most out of the computational side of **Scientific WorkPlace**, this newly revised book offers:

- A solid foundation for doing mathematical calculations using **Scientific WorkPlace** (Chapters 1 and 2)

- Instructions for using **Scientific WorkPlace** to do precalculus, calculus, linear algebra, vector analysis, differential equations, statistics, and applied modern algebra (Chapters 1–9)

- New and revised information on creating plots (Chapters 3–4, 6–7)

- Expanded information on document preparation that will help you create professional-quality documents using the internationally accepted mathematics typesetting standard, $T_EX$ (Appendix A)

**"...a comprehensive, thoughtful, and thought-provoking guide to one of the most exciting developments in the teaching of mathematics: the blend of mathematical software with a rich and natural textual interface."**

Manfred E. Szabo
Concordia University

---

**ORDER FORM**

To order, fill out this coupon and return it to Brooks/Cole along with your check, money order, or credit card information.

**Yes! Please send me a copy of Doing Mathematics with Scientific WorkPlace by Darel Hardy and Carol Walker**

_____ copies (ISBN: 0-534-34049-0) @ $20.75 each.
(A savings of 20% off the regular list price of $25.95!)                    Subtotal   _____

(Residents of AL, AZ, CA, CO, CT, FL, GA, IL, IN, KS, KY, LA, MA, MD, MI, MN, MO, NC, NJ, NY, OH, PA, RI, SC, TN, TX, UT, VA, WA, WI must add appropriate sales tax.)   Tax   _____

Payment Options                                                              Handling   _____
_____ Purchase order enclosed

_____ Check or Money Order enclosed                                        Total   _____

_____ Charge my _____ VISA _____ MasterCard _____ American Express

Card Number _____ Expiration Date _____

Signature_____

Please ship to: (Billing and shipping address must be the same.)

Name_____

Institution _____

Street Address _____

City _____ State_____ Zip+4_____

Telephone ( _____ ) _____

**Mail to: Brooks/Cole Publishing Co., Dept. Doing Math Offer** 511 Forest Lodge Road, Pacific Grove, CA 93950-5098
**Phone: (408) 373-0728 ext. 233 or Fax: (408) 375-6414** E-mail: adrienne_carter@brookscole.com
Visit Brooks/Cole on the World Wide Web: http://www.brookscole.com/brookscole.html or ftp.brookscole.com
Prices subject to change without notice.

*SECURE WITH TAPE*

NO POSTAGE
NECESSARY
IF MAILED
IN THE
UNITED STATES

# BUSINESS REPLY MAIL
FIRST CLASS          PERMIT NO. 358          PACIFIC GROVE, CA

POSTAGE WILL BE PAID BY ADDRESSEE

ATT:  MARKETING

**Brooks/Cole Publishing Company
511 Forest Lodge Road
Pacific Grove, California 93950-9968**